Study Guide

to accompany

Ecology and Field Biology

Fifth Edition

Robert Leo Smith
West Virginia University

An Imprint of Addison Wesley Longman, Inc.

Menlo Park, California • Reading, Massachusetts • New York • Harlow, England
Don Mills, Ontario • Sydney • Mexico City • Madrid • Amsterdam
Bonn • Paris • Milan • Singapore • Tokyo • Seoul • Taipei

ISBN 0-065-00978-9

1 2 3 4 5 6 7 8 9 10–CRS–00 99 98 97 96

The Benjamin/Cummings Publishing Company, Inc.
2725 Sand Hill Road
Menlo Park, California 94025

TABLE OF CONTENTS

TO THE STUDENT

Ecology and Field Biology, like other relatively large texts, may seem a little overwhelming. Because ecology is such a diverse subject, an introductory text must provide a sweep of the subject. If your ecology course is just a semester long, instructors will emphasize different parts of the book. This fact does not mean you should ignore or overlook the rest of the text. Topics of one chapter relate to other parts of the text. Related topics are indicted by cross references given at the end of each chapter. Your instructor may refer you to some of the related material.

The purpose of this guide is to help you master the subject. To get the most out of the text and this guide you should first read the chapter outline and chapter concepts. Then turn to this guide and read through the learning objectives. Next read the chapter, using the study questions in this guide as well as the review questions at the end of each chapter. Take notes as necessary. Note taking is done best by writing answers to the study questions. Review the summary, then turn to the chapter guide to complete your mastery of the subject.

The study guide consists of :

Chapter Outlines. The outlines come from the text. They serve as a reference to the material being covered.

Summary. This section prompts you to read the summary of the chapter in the text. It also alerts you to related material in other chapters that will add more examples or applications of the topics in the chapter. For example Chapter 5 discusses the physical properties of water. In the Summary section you will find listed page references to additional related material on influence of those physical properties to life in lakes, streams and rivers, and marine environment. The cross references tie what seems to be disparate topics into a related whole. Remember, in ecology everything is tied to something else.

Study Questions. This section is an important part of the study guide. The questions steer you through the chapter by emphasizing the important points. I have passed out lists of similar questions to my classes for a number of years. My students found them very useful. In fact when I asked a recent class what items would be most useful to them in a study guide, the students said a list of study questions. They found that answering these questions was a most valuable means of mastering the material in the chapter.

Key Terms and Phrases. A common impression students gain from science texts, especially biology texts, is an overemphasis on terminology, often derided as jargon. Many students believe that passing a course requires memorization of and testing on terminology. Such memorization should not be the end-all of the course. Concepts are more important than terminology. Unfortunately, understanding concepts requires some familiarity with terminology. The important terms in each chapter are given in the lists.

Review of Key Terms. This section of fill-in questions tests your knowledge of the terms. As you will discover, most of the questions emphasize your ability to apply the term correctly rather than simply recognize the definition.

Self Test. The self tests enable you to discover how well you have mastered the material in the chapter. The questions are not comprehensive, but they are varied enough for you to discover what areas need more review. Self test consists of three types of questions: *true-false*, *matching*, and *short answer*, but in some chapters you will encounter some *greater than*, *lesser than contrast* questions and *labeling* questions. The short answer questions are designed to test your ability to apply the information gained from the chapter. At the end of Chapters 20 and 27 you will find a Comprehensive Application Exercise. The exercises, based on narrative of an imaginary field situation, test your ability to apply concepts covered in preceding chapters. The Exercise at the end of Chapter 20 is based on Chapters 22 through 27. You will notice an absence of multiple choice questions. I have an aversion to them. Typically, to make a sufficient number of choices for a question, nonsense or wholly unrelated choices are included in the answer list, simply as filler to provide the four or five choices necessary. Other choices are ambiguous or at best debatable. Completing these self tests should give you a high degree of confidence in taking classroom tests.

Answers. You will find answers to the questions at the end of each chapter. Don't look until you have completed your tests.

Good luck.

PART 1

INTRODUCTION

CHAPTER 1
ECOLOGY: ITS MEANING AND SCOPE

Chapter Outline

Ecology Defined
The Development of Ecology
   Plant Ecology
   Animal Ecology
   Physiological Ecology
   Population Ecology
   Ecosystem Ecology
   Cooperative Studies

Tensions within Ecology
   Plant vs. Animal Ecology
   Organismal vs. Individualist Ecology
   Holism vs. Reductionism
   Theoretical vs. Applied
Summary

Learning Objectives

After completing this chapter you should be able to:
- Define ecology.
- Discuss the evolution of ecology as a science.
- Discuss the several approach current approaches to ecology.
- Explain the relationship of ecology to current environmental problems.

Summary

After reading this chapter and before continuing with the material below, read the Chapter Summary on pages 12-13. This chapter deals mostly with a brief history and evolution of ecology. Although many of you probably have little interest in history, an appreciation of the development of ecology is essential to understanding modern ecology.

Study Questions

1. What is ecology? (4)
2. Why was plant geography a stimulus for development of modern ecology? (5)
3. What are some different fields of ecology? How did each develop? (6-8)
4. What differences separate organismal concept of ecology from the individualistic concept? (9)
5. How do these concepts relate to holism and reductionism in ecology? (9-10)
6. What is applied ecology? What is its relationship to theoretical and ecosystem ecology? (10-11)
7. Discuss the relationship of ecology to current environmental problems. (11)
8. Why do current environmental problems and conflicts involve more than just the science of ecology? (11-12)

Key Terms and Phrases

ecology
ethology
behavioral ecology
physiological ecology
chemical ecology
population ecology
population genetics

evolutionary ecology
limnology
ecosystem ecology
systems ecology
emergent theory
ecosystem

holistic ecology
reductionist ecology
applied ecology
conservation biology
landscape ecology
restoration ecology

1. _____ is the study of the structure and function of nature.

2. _____ _____ is concerned with population growth, regulation, and interactions, whereas _____ _____ is concerned with the genetics of and evolution in populations.

3. The study of the interaction of population dynamics, genetics, natural selection, and evolution is _____ _____.

4. An _____ consist of a biotic community interacting with the abiotic environment.

5. _____ _____ states that natural association have certain unique properties that arose from lower levels of organization.

6. The study of life in the open water of lakes is _____.

7. The use of ecological theory and models in ecosystem and natural resource management is _____ _____.

8. _____ _____ is concerned with spatial distribution of flows between ecosystems.

9. Application of experimental research and ecological principles to restoration of disturbed lands is _____ _____.

10. Ecosystem ecology is basically _____, whereas population ecology is _____.

11. _____ _____ is concerned with habitat destruction and fragmentation and effects of population reduction.

12. _____ _____ is concerned with the interaction of animals with their living and nonliving environments with an emphasis on natural selection. It developed from _____, the study of evolution and function of behavior.

13. Ecologists studying the use of natural substances by plants and animals in species recognition, reproduction, and defense are involved in _____ _____.

14. Study of the responses of organisms to temperature, moisture, light, and other environmental factors is _____ _____, or _____.

Self Test

True and False

1. _____ Reductionists state that in the study of complex systems the sum is greater than the parts.

2. _____ Ecosystem ecologists can study only a part of the ecosystem at any one time.

3. _____ Description and classification of plant communities was the major interest of early American plant ecologists.

4. _____ Ecological principles are accepted by developers and politicians.

5. _____ The major difference between reductionists and holists is that the latter study larger parts.

Matching

Match the name on the left with the identifying attribute on the right.

1. _____ Darwin   A. Interrelation between plants and animals--bioecology.

2. _____ Tinbergen   B. English animal ecologist--theory of the niche.

3. _____ Malthus   C. Brought environmental problems to public attention

4. _____ Elton   D. Concept of plant succession.

5. _____ Clements   E. Relation of population growth and food supply.

6. _____ Lindeman   F. Theory of natural selection.

7. _____ Shelford   G. Behavior: ethology

8. _____ Carson   H. Trophic dynamic aspects of ecology.

Short Answer Question

What is the relationship of applied ecology to ecosystem and theoretical ecology?

ANSWERS

Key Term Review

1. ecology
2. population ecology, population genetics
3. evolutionary ecology
4. ecosystem
5. emergence theory
6. limnology
7. applied ecology
8. landscape ecology
9. restoration ecology
10. holistic, reductionist
11. conservation biology
12. behavioral ecology, ethology
13. chemical ecology
14. physiological ecology, ecophysiology

Self Test

True and False

1. F
2. T
3. F
4. F
5. T

Matching

1. F
2. G
3. E
4. B
5. D
6. H
7. A
8. C

Short Answer Question

1. Applied ecology relates to forestry, range, and wildlife management, that is ecosystem management, as well as populations of organisms. For that reason it must apply principles of ecosystem ecology and population ecology to achieve its goals.

CHAPTER 2
EXPERIMENTATION AND MODELS

Chapter Outline

Inductive and Deductive Approaches

Collecting Data

Testing Hypotheses
  Types of Errors
  Statistical Power Tests

Models and Predictions
  Statistical Models
  Nonstatistical Models

Validation

Summary

Learning Objectives

After completing this chapter you should be able to:
- Define the meaning of and construction of a hypothesis.
- Distinguish between inductive and deductive approaches to testing a hypothesis.
- Distinguish between the dependent and independent variables.
- Explain what is involved in the testing of hypotheses.
- Distinguish between Type I and Type II errors.
- Explain the importance of the statistical power test.
- Explain the role of models in ecology
- Distinguish between the different kinds of models.
- Explain the importance of model validation.

Summary

After reading the chapter and before proceeding with the material below read the Chapter Summary on pages 24-25. Because this chapter deals marginally with statistics, also refer to Appendix A, pages 688-689.

Study Questions

1. What is a hypothesis? (16)
2. What is the difference between an independent variable and a dependent variable? (16)
3. Distinguish between the inductive and deductive methods of testing hypotheses. (16)
4. Why are replicates and controls important in ecological studies? (17)
5. Why can pseudoreplication become a problem in ecological studies? (17-18)
6. What are null and alternative hypotheses? (18)
7. What are Type I and Type II errors an experimenter can make in accepting or rejecting a hypothesis? (19-20)
8. What is a statistical power test and what is its purpose? (20-21)
9. Define a model. Of what value are models in ecology? (21)
10. What are the limitations of statistical models in ecology? (21)

11. Distinguish between an analytical model and a simulation model. (21-22)
12. What is meant by validation of a model? How can models be validated? (23-24)

Key Terms and Phrases:

hypothesis
null hypothesis
inductive method
deductive method
dependent variable
independent variable
controls

replicates
pseudoreplication
variance
standard deviation
Type I error
Type II error
statistical power test

model
statistical model
nonstatistical model
analytical model
simulation model
validation
parameter

Review of Key Terms

1. A statement about an observation that can be tested experimentally is a
   _____.

2. An abstract representation of a real system is a _____.

3. Rejection of null hypothesis when it is true is a _____ Error.

4. Rejection of alternative hypothesis when it is true is a _____ Error.

5. A _____ model predicts the value of a dependent variable based
   on mathematical functions for which parameters have been empirically derived.

6. A _____ _____ _____ measures the probability of rejection of
   null hypothesis (no treatment effect) at a particular alpha level when the null
   hypothesis is false.

7. In _____ models, only one solution to a problem exists for a given set
   of parameters.

8. _____ models cannot be solved analytically; they require use of a
   computer.

9. _____ is an objective test of how much confidence we can place in
   the predictive abilities of the model.

10. The measurements squared of the position of each observation in a set of
    observation relative to the mean of the set is the _____.

11. The _____ _____ expresses the variation of observations
    about a mean in the same units as the raw data.

12. Experimental plots receiving the same treatments are called _____.

13. Plots receiving no treatment are the _____.

8

14. Situations in which the observable units or plots may have an influence on each other is called _____.

15. Hypothesis are tested statistically by creating a _____ hypothesis that states that no difference exists between control and treatment units.

16. The variable a researcher allows to change is the _____ variable.

17. The _____ variable responds to changes in another variable.

18. The _____ method involves the gathering of empirical data to arrive at a generalization.

19. Any numerical value describing a characteristic of a population is called a _____.

Self-Test Questions

Short Answer Questions

1. Below is a labeled graph. Indicate which axis is the independent variable and which is is the dependent variable by placing an X on the independent variable axis and a Y on the dependent variable axis. What does the graph tell you?

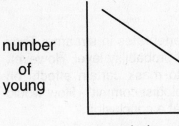

number
of
young

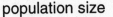

population size

2. Indicate which of the following is an analytical model and which is a simulation model. Give a reason for your choice.

_____ **a.** $dN/dt = rN(K-N)/N$ where N is initial population size, r is the rate of growth, K is the carrying capacity of the habitat, and $dN//dt$ is the population size at time t.

_____ **b.** $V = Fv$ where V = a compartment system, v = value of a system variable, and F = a transfer function between compartments.

3. Indicate which of the following represents an inductive and deductive approach to testing hypotheses:

a. A biologist notice that hummingbirds preferred red-colored flowers in the garden and rarely visited white and yellow flowers. The biologist surveyed the feeding behavior of ruby-throated hummingbirds over a large region and the information gathered indicated that hummingbirds showed a decided preference for red flowers. The biologist concluded that hummingbirds preferred red-colored flowers (and feeding tubes).

b. A biologist presented hummingbirds with an array of different colored feeders at a number of locations, and recorded the frequency of their visits to each. Analysis of data collected indicated that the hummingbirds showed a decided preference for red feeding tubes over yellow, black and white feeding tubes.

4. The decline in amphibians, especially certain frogs and salamanders, has been attributed to acidification of their habitats by acid deposition. A biologist hypothesized that low pH decreases the survival and larval development of cricket fogs in her region, and is going to undertake an experimental study to test that hypothesis. Write a testable null hypothesis and an alternative hypothesis or this study.

a. H_o

b. H_a

5. Two ecologists studying the effects of Chemical X on invertebrates in streams. They accepted the null hypothesis of no effects at a 95 percent probability level. However, their study. unknown to them, was sufficiently biased, to mask certain effects on some aquatic invertebrates. What type of error did the ecologist commit? How would a statistical power test have helped these ecologist arrive at a conclusion.?

ANSWERS

Key Term Review

1. hypothesis
2. model
3. Type I
4. Type II
5. statistical
6. statistical power test
7. analytical
8. simulation
9. validation
10. variance

11. standard deviation
12. replicates
13. controls
14. pseudoreplication
15. null
16. independent
17. dependent
18. inductive
19. parameter

Self Test

Short Answer Questions

1. The graph tells us that as the population size increases, the number of young decreases.

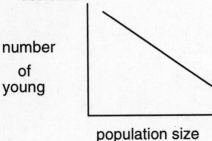

number
of
young

population size

Number of young = Y axis
Population size = X axis

2. **a.** The model is analytical. There are three specific values to be inserted in the model. inserted in the model: the rate of growth, carrying capacity, and the initial population. Growth of the initial population is defined by the given carrying capacity and the given rate of growth. Given those parameters, the model has one solution.

   b. This is a simulation model. In spite of its simplicity of the model of a system, it involves many parameters. It involves a number of subcompartments with rate of flow between them. The output from one subcompartment can influence the function of another. The complexity of rate changes and changes in subcompartments require computer to handle this model. Change in any one given variable or forcing functions changes all other parameters.

3. **a.** Deductive. The biologist observed that the hummingbirds preferred red flowers and red feeding tubes. The biologist investigated known facts and observations and deduced from these that the hummingbirds preferred the color red. He arrives at a specific conclusion based on a general body of facts and observations. The biologist when from general to the specific, took a deductive approach.

    b. Inductive. The biologists established a hypothesis, gathered empirical data based on experiments, repeated the experiments at different locations, and arrived at a general conclusion. The biologist went from the specific to a general conclusion: that hummingbirds preferred red colored flowers and feeding tubes.

4. H_o: Low pH has no effect on survival and larval development of cricket frogs.

    H_a: Low pH decreases the survival and larval development of cricket frogs.

5. The experimenters committed a Type II error. They failed to reject a false null hypothesis. Because hypotheses are statistical, based on probability, they should have used a statistical power test to determine the probability of rejection of null hypotheses at a particular alpha level.

PART 2

THE ORGANISM AND ITS ENVIRONMENT

CHAPTER 3
ADAPTATION

Chapter Outline

The Meaning of Adaptation   Homeostasis
Tolerance   Summary

Learning Objectives

After completing this chapter you should be able to:
- Discuss the meaning of adaptation.
- Define fitness and natural selection.
- Distinguish between ecotype and phenotype.
- Explain the law of minimum, the law of limiting factors, and the law of tolerance.
- Define acclimatization and note its relationship to the law of tolerance.
- Distinguish between poikilotherms and homeotherms.
- Discuss homeostasis and the role of positive and negative feedback.

Summary

After reading this chapter and before proceeding with the material below, read the Chapter Summary on page 34. You will find related material on the following pages: plant response to moisture, 70-75; animal response to moisture, 75-77; plant responses to temperature, 83-85; animal responses to temperature, 85-95; plant adaptations to light intensity, 102-104; photoperiodism, 106-110; seasonality, 110-112; nutrients and plants, 112-124; tundra, 250-261.

Study Questions

1. Define adaptation. (30)
2. How do organisms become adapted to their environment? (30)
3. What is fitness? How does fitness relate to natural selection? (30)
4. Contrast the law of the minimum with the law of limiting factors. (31)
5. What are ecotypes? (30)
6. What is a phenotype? (30)
7. What is phenotypic plasticity and how does it relate to environmental conditions? (30-31)
8. What is the law of tolerance? (31)
9. Relate acclimatization to the law of tolerance. (32)
10. What is homeostasis; homeostatic plateau? (33), (34)
11. How do positive feedback and negative feedback function in the maintenance of homeostasis? (33)

Key Terms and Phrases

adaptation	phenotype	poikilotherm
fitness	law of minimum	homeotherm
natural selection	law of limiting factors	homeostasis
ecotype	law of tolerance	set point
genotype	acclimatization	positive feedback
set point	homeostatic plateau	

Key Term Review

1. Genetic characteristics of an organisms is its _____. The physical expression of genetic characteristics is the organism's _____.

2. Populations of plants adapted to local environmental conditions are _____.

3. The contribution an individual makes to future generations is its _____.

4. The _____ ___ _____ states that the well being and growth of an organism is limited by the needed resource in shortest supply.

5. The _____ ___ _____ _____ states that environmental conditions can limit survival, growth, and reproductive success.

6. _____ are organisms that maintain a relatively constant body temperature in the face of a varying environment, whereas _____ more or less follow external conditions.

7. The _____ ____ _____ states that too much or too little of a needed resource will limit the response of an organism.

8. The floating leaves of pondweed are broad, whereas the leaves submerged in water are ribbonlike. This change in leaf structure brought about by different environmental conditions is called _____ _____.

9. _____ _____ favors any heritable or behavioral characteristics that increases _____.

10. Gradual adjustment of fish to changing water temperature is _____.

11. Maintenance of relatively constant internal environment in varying environmental conditions is _____.

12. The limited range of tolerances of an organism is its _____ _____.

13. Movement away from a _____ _____ is _____
_____. the response that halts or reverses this movement away is
_____ _____.

Self Test

True and false

1. _____ Positive feedback moves away from the set point.

2. _____ Homeostasis is maintained by negative feedback.

3. _____ The range of tolerance is fixed.

4. _____ Acclimatization is an example of a plastic, short-term response to environmental change.

5. _____ Poikilothermic maintain temperature homeostasis by an internal feedback mechanism.

Matching

Match the following statements on the left with the term on the right.

1. _____ Heritable traits that maintain fitness.   A. ecotype

2. _____ Physical expression of genotype.   B. fitness

3. _____ Individual's contribution to future generations.   C. natural selection

4. _____ Differential reproductive success.   D. adaptation

5. _____ Adaptation to local conditions.   E. phenotype.

Short Answer Question

1. Individual A and individual B of the same species produce 9 and 5 offspring respectively. Of the 9 offspring of A, 2 successfully reproduce. Of the 5 offspring of B, 3 successfully reproduce. Which one, A or B, has the greatest fitness? Why?

ANSWERS

Key Term Review

1. genotype, phenotype
2. ecotypes
3. fitness
4. law of minimum
5. law of limiting factors
6. Homeotherms, poikilotherms
7. law of tolerance
8. phenotypic plasticity
9. natural selection, fitness
10. acclimatization
11. homeostasis
12. homeostatic plateau
13. setpoint, positive feedback, negative feedback

Self Test

True and False

1. T
2. T
3. F
4. T
5. F

Matching

1. D
2. E
3. B
4. C
5. A

Short Answer Question

1. Individual B has the greatest fitness. Although A produced 9 offspring, only 2 of them reproduced, whereas 3 of B's 5 offspring reproduced. B contributed more to future reproduction than did A.

CHAPTER 4
CLIMATE

Learning Objectives

After completing this chapter you should be able to:

- Distinguish between weather and climate.
- Describe how climate controls prevailing abiotic conditions.
- Describe the fate of solar energy when it reaches Earth.
- Explain how the atmosphere warms and how this warming influences global air movements.
- Explain how adiabatic heating and cooling influence regulate air movements.
- Compare and contrast stable and unstable air masses and inversions.
- Describe microclimates and state their role in determining local environments.
- Discuss the role of climate in the global distribution of vegetation.

Summary. After reading the chapter and before proceeding with the material below read the Chapter Summary on pages 60-02. You will find related material and examples on the following pages: plant responses to moisture, 70-75; animal responses to moisture, 75-77; thermal energy exchange, 80-82; plant responses to temperature, 80-83; animal responses to temperature, 85-95; seasonality, 110; nature of energy, 168-169; greenhouse effect. 205-206; grasslands, 226-236; savanna, 236-239; shrubland, 239-246; deserts, 246-250; forests, 264-288; tundra, 250-261.

Study Questions

1. What is the major determinant of climate? What is meant by the solar constant? (36)
2. What wavelengths of solar radiation are filtered out by the atmosphere? What wavelengths reach the Earth in quantity? (36, 38)
3. What is albedo and how does it relate to solar radiation? (37)
4. What is an adiabatic process? Dry adiabatic lapse rate? Moist adiabatic lapse rate? (38-39)
5. What are the major differences between stable and unstable air masses? (38-39)
6. What is the Coriolis force? How does it influence global air circulation? (39)
7. Why are lower latitudes heated more than polar ones? What effect does this unequal heating have on atmospheric circulation over Earth? (40-42)
8. What is the intertropical convergence? What is its ecological significance? (41)
9. What produces ocean currents? What is a gyre? (40-42)
10. Define humidity. What is the difference between absolute humidity and relative humidity? (45)
11. What is an inversion? What is unique about it, temperature-wise? (46)
12. What is a subsidence inversion? A marine inversion? (47-48)
13. How does vegetation influence microclimate? Why is the ground layer within vegetation a favorite microclimate for insects and small mammals? (51-52)
14. How do soil properties influence microclimate? (52)
15. What are the major microclimatic differences between a north-facing and a south facing slope? (48)
16. How do these differences influence vegetation? (48)
17. What is a frost pocket? A heat island? (53)
18. Contrast the climate of city with that of the countryside. (53-54)
19. How do temperature and rainfall patterns influence vegetation? (55-56)
20. What are the broad vegetation zones in North America? What influences this zonation? (55-56)
21. On what is the Holdridge vegetation classification system based? (56-57)
22. What is an ecoregion? What is the use of this classification? (59-60)

Key Terms and Phrases

weather	Coriolis effect	relative humidity
climate	intertropical convergence	inversion
solar constant	anticyclones	subsidence inversion
albedo	gyre	marine inversion
adiabatic process	upwelling	microclimate
dry adiabatic lapse rate	El Nino	frost pocket
moist adiabatic lapse rate	rain shadow	heat island
dew point	climograph	life zone
stable air mass	evapotranspiration	ecocline
unstable air mass	vapor pressure	ecoregions
vapor pressure deficit	biomes	

Key Term Review

1. _____ refers to temperature, humidity, winds, and other conditions at a given time and place.

2. The summation of weather conditions over a long period is _____.

3. Cal/cm^2 is the _____ _____.

4. An _____ is an atmospheric condition in which temperature of the air mass increases with height.

5. The _____ _____ acts on the north-south global air currents to give them a east-west component.

6. The great circular water motions of the Atlantic and Pacific Oceans are called _____.

7. _____ is the percentage of solar radiation reflected by Earth.

8. _____ is the amount of water a parcel of air can hold at a given temperature.

9. In an _____ _____ expansion of an air mass results in cooling and compression in heating.

10. The temperature at which moisture in the atmosphere condenses is the _____ _____.

11. The rapidity with which temperature decreases with altitude is the _____ _____ _____.

12. The vertical motion of water in the ocean by which cold. dense subsurface water moves to the surface of the ocean is known as _____.

13. The discharge of water from Earth's surface to the atmosphere by evaporation from bodies of water and other surfaces and by transpiration by plants is known as _____.

14. An atmospheric high pressure circulation with clockwise rotation in the Northern Hemisphere and counterclockwise in the Southern Hemisphere, and undefined at the Equator is known as an _____.

15. A climatically defined region characterized by similarity in soil, in plant and animal life, and environmental adaptations is a _____ _____.

16. The largest recognizable biogeographical communities broadly corresponding to climatic regions are _____.

17. Atmospheric conditions prevailing in a relatively small space, usually in a layer near the ground make up _____.

18. A gradation from one community to another along an environmental gradient such as temperature or moisture is an _____.

19. Temperatures near and within a city are much higher than the surrounding country side and create a _____ _____.

20. Drier environmental conditions exist on the lee side of mountains because the lee side lies within a _____ _____.

21. A sequential plot of the average values of monthly rainfall against temperatures producing a visual description of climate of a region is a _____.

22. This low latitudinal zone meeting of air masses at the boundary of northeasterly and southeasterly trade wind form low latitudinal depressions that move latitudinally with the seasons known as _____ _____.

23. Gradual lowering of a layer of air over a wide area is a _____ _____.

24. A smoke column that drifts apart after a limited rise indicates a _____. _____ _____.

25. The _____ system of ecosystem classification is based on a geographical area with similar soil, climate, topography and vegetation.

26. Clouds growing vertically and smoke rising to great height indicate an _____ _____ _____.

27. Cool mist air from ocean flowing over low-lying land is topped by a warmer drier air mass to create a _____ _____.

28. Amount of pressure water vapor exerts independent of dry air is _____ _____. The pressure water exerts when the air is calm is _____ _____ _____. The difference between these two at a given temperature represents _____ _____ _____.

Self-Test

True and False

1. _____ As a result of the rain shadow effect, mesic plants grow on the lee side of the mountain.

2. _____ Smoke rising straight up into the air and spreading out means atmospheric unstability.

3. _____ A rising unstable air mass cools at a rate faster than adiabatic.

4. _____ Gyres, large circular currents in the oceans, rotate to the left in the Northern Hemisphere.

5. _____ A nighttime inversion results when the air near the ground cools by radiation and becomes blocked by warmer air above.

6. _____ The Coriolis force deflects air and water movements and prevents a simple flow from equator to the poles.

7. _____ The active surface receives the full impact of solar radiation.

8. _____ When gases in the atmosphere expand, they heat.

9. _____ During temperature inversions, air pollution increases..

10. _____ Long-wave radiation easily passes through the atmosphere and escapes from Earth.

11. _____ Vegetation alters wind movement, moisture, and soil temperature to create microclimates.

12. _____ North-facing slopes receive more sunlight than south-facing slopes.

Matching

Ecologists recognize that the vegetation of a region is determined by an interaction of temperature and rainfall. Match the abiotic conditions with expected vegetation type::

A. Tundra   B. Tropical rain forest   C. Savanna
D. Grasslands   E. Desert   F. Temperate forests

1. _____ High temperature, low precipitation

2. _____ High temperature, high precipitation

3. _____ High temperature, moderate seasonal rainfall

4. _____ Moderate temperature, moderate to low precipitation

5. _____ Cold temperature, low precipitation

6. _____ Moderate temperature, moderate precipitation

Select the appropriate terms from the list below to match the following statements.

A. Inversion   B. Dew point   C. Ecosystem
D. Biome   E. Ecoregion   F. Climate
G. Weather   H. Microclimate   I. Life Zone

7. _____ Broad natural biotic units.

8. _____ Ecosystem classification based on mean annual precipitation and mean annual biotemperature.

9. _____ Temperature at which atmospheric water condenses.

10. _____ Atmospheric condition in which the temperature of the air increases with height.

11. _____ Defines the condition under which most organisms live.

12. _____ Product of weather over time.

Short Answer Questions

1. Given that the sun illuminates all parts of Earth during part of the 24-hour day, why is the temperature at and near the equator so much higher than that at the poles?

2. What are the two major climatic differences between a north-facing and south-facing slope?

On a clear day in early fall you noted that the temperature on the ridgetop of a small mountain was 75° F and in the valley, 85° F. In the evening you felt a cool breeze flowing downslope. The mountaintop temperature by night had dropped to 65° and in the valley to 55° F. By morning fog filled the valley, but leveled of just below the mountain ridge.

3. What caused the downslope evening breeze?

4. What was happening to the air mass in the valley?

5. Why did fog fill only the valley?

ANSWERS

Key Term Review

1. weather
2. climate
3. solar constant
4. inversion
5. Coriolis force
6. gyre
7. albedo
8. relative humidity
9. adiabatic process
10. dew point
11. adiabatic lapse rate
12. upwelling
13. evapotranspiration
14. anticyclone
15. life zone
16. biome
17. microclimate
18. ecocline
19. heat island
20. rain shadow
21. climograph
22. intertropical convergence
23. subsidence inversion
24. stable
25. ecoregion
26. unstable
27. marine inversion
28. vapor pressure, saturation vapor pressure. vapor pressure deficit

Self-Test Questions

True and False

1. F	7. T
2. F	8. F
3. F	9. T
4. F	10. F
5. T	11. T
6. T	12. T

Matching

1. E	7. D
2. B	8. E
3. C	9. B
4. D	10. A
5. A	11. H
6. F	12. G

Short Answer Questions

1. At the equator, solar radiation is direct. At the poles Earth intercepts solar radiation at an angle and have to penetrate a deep blanket of air. Thus more energy is scattered back to the atmosphere.

2. North-facing slopes are cooler and more moist. South-facing slopes are warmer and drier.

3. Warm lighter air rises; cold air sinks. Cooling air on the upper slopes becomes more dense and flows down slope.

4. Air mass near the ground in the valley was experiencing radiation cooling and cool air was draining into the valley from the upper slopes.

5. Cool nighttime air was held in the valley by an inversion. When its temperature reached the dew point, water in the air condensed as fog. The upper level of the fog marks the height of the nighttime inversion.

CHAPTER 5
WATER BALANCE

Learning Objectives

After completing this chapter you should be able to:
- Describe the structure of a water molecule.
- State the several properties of the water molecule that make it so important to life.
- Describe the distribution of water on Earth.
- Explain the water cycle.
- Compare and contrast the global water cycle with a local water cycle.
- Summarize the adaptive strategies of plants to moisture deficits and excesses.
- Describe the adaptive strategies of animals to different moisture conditions.
- Compare and contrast drought resistance to drought tolerance.
- Relate the availability of or lack of moisture to the distribution of vegetation.

Summary

After reading the chapter and before continuing with the material below read the Chapter Summary pages 77-78. You will find related material in the following pages: adaptation, 30-31; evaporation, 44; humidity, 45; climate and vegetation, 54-60; soil moisture, 68,133; photosynthesis, 154-157; lakes and ponds, 290-301; streams, 301-313; freshwater wetlands, 313-323; salt marsh, 348-357; desert, 246-250.

Study Questions

1. What unique features of the water molecule enable water to assume three forms. (64-65)
2. Describe the following physical features of water and the ecological importance of each: specific heat, latent heat, cohesion, viscosity, and surface tension. (65-67)
3. What is the ratio of salt water to fresh in the global distribution of water? Where is most of the fresh water located? (66)
4. Describe the local water cycle, showing the relationship among precipitation, evaporation, interception, throughfall, stemflow, infiltration, and percolation?

(66-68)
5. Contrast the precipitation--evaporation budget over the oceans with the same budget over land. (69-70)
6. Explain the processes by which plants move water from soil to leaves. (70-71)
7. Distinguish between a poikilohydric plant and a homiohydric plant. Give an example of each. (71)
8. In what general ways do plants respond to drought? (71-72)
9. In what ways have plants of semiarid regions adapted to aridity? (72)
10. What are phreatophytes? Halophytes? (72)
11. How do plants adapt to saline environments? (72)
12. What problems are created for plants by flooding and how do they respond? (73-75)
13. How do some animals adapt to moisture stress in the desert? In a saline environment? (75)
14. What physiological mechanisms enable animals of the desert to tolerate a lack of moisture? (76)

Key Terms and Phrases

hydrogen bonding	evaporation	field capacity
specific heat	capillary water	aerenchyma
latent heat	heat of fusion	pneumatophores
surface tension	osmotic pressure	hyperthermia
viscosity	turgor pressure	interception
homeohydric	stemflow	poikilohydric
water potential	throughflow	halophyte
water use efficiency	infiltration	preatophyte
transpiration	osmotic potential	

Key Term Review

1. The linkage between water molecules is know as _____ _____.

2. It takes considerable amount of energy to convert liquid water to water vapor because water has a high _____ _____ _____.

3. _____ _____ is a measure of the energy of water determined by two opposing drives, osmotic pressure and turgor pressure.

4. The resistance between layers of water is called _____.

5. At the surface of a body of water, _____ bonds pull on the molecules at the surface and create a high _____ _____.

6. If precipitation intercepted by a forest canopy exceeds its storage capacity, then water drips off the leaves as _____ and runs down the trunk as _____.

7. Considerable energy is needed to raise the temperature of water because water has _____ _____ _____.

8. Water must lose a considerable amount of heat to turn to ice because water has a
   _____ ____ _____ .

9. _____ have long tap roots that allow plants to use water deep
   beneath the ground's surface.

10. _____ are plants adapted to live in saline environments.

11. Pumping salts against a _____ gradient is one type of
    _____ _____ _____ .

12. Water held by capillary forces between soil particles is _____ _____ .
    The maximum amount of water held in the soil at one-third atmosphere of pressure
    after gravitational water has drained away is _____ _____ .

13. _____ _____ _____ refers to the amount of water a plant
    loses during the fixation of one mole of CO_2.

Self Test

Matching

Match the lettered characteristics of water with numbered statements below. You may
use a lettered response more than once.

A. Surface tension   D. Viscosity   G. Heat of fusion
B. High specific heat   E. Polar molecule
C. Heat of evaporation   F. Water is less dense as a solid than a liquid

1. _____ A water strider runs across the surface of the water.

2. _____ The water along the banks of a stream moves very slowly, while the water
             in the center of the stream moves rapidly.

3. _____ Considerable energy is needed to raise the temperature of water.

4. _____ Water is an excellent solvent ; many substances will dissolve in it.

5. _____ Considerable heat must be applied to a pan of liquid water to turn it to
             vapor.

6. _____ Water must lose considerable heat before it freezes.

7. _____ Water is used as a coolant in a power plant.

8. _____ Fish can live at the near the bottom of a lake in winter.

9 _____ The climate of land near a large body of water is warmer in winter and
            cooler in summer than land area further inland.

True and False

1. _____ The attraction of one water molecule to another is due to the polarity of the water molecule.

2. _____ Evaporation is the physical process of converting liquid water into vapor.

3. _____ Water moves in the direction of high water potential.

4. _____ Globally, precipitation over land is greater than precipitation over the ocean.

5. _____ Plants experience ill-effects from flooding because no gas exchange can take place between the roots and the soil.

6. _____ Poikilohydric plants maintain a relatively stable water balance in their tissues.

7. _____ Plants experiencing flooding show many of the same visible symptoms as plants experiencing drought conditions.

8. _____ Precipitation is the driving force of the water cycle.

Short Answer Questions

1. In what way is a salt marsh a desert environment for plants?

2. As you are traveling along an interstate, you notice that trees standing a pool of water created by a road fill are dead. Explain why these trees died.

3. How does a plant draw water from the soil to the leaves?

ANSWERS

Key Term Review

1. hydrogen bond
2. heat of evaporation
3. water potential
4. viscosity
5. hydrogen, surface tension
6. throughfall, stemflow
7. high specific heat
8. heat of fusion
9. phreatophytes
10. halophytes
11. concentration, active transport
12. capillary water, field capacity
13. water use efficiency

Self Test

True and False

1. T
2. T
3. T
4. F
5. T
6. F
7. T
8. F

Matching

1. A
2. B
3. B
4. E
5. B
6. G
7. B
8. F
9. B

Short Answer Questions

1. Although salt marsh plants have their roots in a wet soil, they still physiologically experience "drought" or "arid" conditions because the salinity of the substrate limits the amount of water they can absorb. To take in water, some plants have to maintain high internal osmotic pressure in their roots and get rid of salts through salt-secreting glands or remove salts at the root membranes.

2. The road fill backed up water, creating a standing pool of water. Because their roots were deprived of oxygen, the trees could not carry on photosynthesis over a prolonged period of time and died.

3. To get water to their leaves, plants have to pull water from the soil. The pumping mechanism involves a water potential gradient. Because the roots have a higher solute concentration and a lower water potential than the soil, water enters the roots. At the other end the leaves of the plant are surrounded an atmosphere with relatively low water potential. Attracted by the lower water potential of the atmosphere, water in the leaves passes through the stomata to the atmosphere (transpiration). To replace the water lost, the plant pulls water from the xylem; the resulting negative water potential eventually reaches down to the roots. This tension pulls water from the soil, up through the stems to the leaf.

CHAPTER 6
THERMAL BALANCE

Chapter Outline

Thermal Energy Exchange
Temperature and Metabolism
Plant Responses to Temperature
  Heat Stress
  Cold Stress

Animal Response to Temperature
  Poikilotherms
  Homeotherms
  Heterotherms
Temperature and Distribution

Learning Objectives

After completing this chapter you should be able to:
- Describe the means of heat transfer between an organism and its environment.
- Explain the relationship between environmental temperatures and the metabolic process of organisms.
- Tell how plants respond to heat and cold.
- Distinguish among poikilotherms, homeotherms, and heterotherms.
- Explain how each of those three groups respond to changes in their thermal environment.
- Discuss how temperature influences the distribution of organisms.

Summary

After reading the chapter and before continung with the material below read the Chapter Summary chapter on page 97. You will find related material in the following pages: adaptation, 30-31; climate and vegetation, 58-61; drought resistance, 71-72; hyperthermia, 76; photosynthesis, 154-157; desert, 246-250.

Study Questions

1. Describe the following conduction, convection, evaporation, and radiation. What is the role of each in energy exchange between an organism and its environment? (80-81)
2. Under what conditions do organisms lose heat to the environment? Gain heat? (80-81)
3. Describe the thermal relationships of an organism with its environment, using a simple formula. (85)
4. How and why do plants differ from animals in their thermal relationship with the environment? (83-85)

5. What is the difference between chilling tolerance, cold resistance, and freezing tolerance in plants? (85)
6. What parts of a woody plant are the most frost tolerant? (85)
7. What is supercooling? How does it function? (85)
8. What is the difference between endothermy, ectothermy, and heterothermy?(86)
9. What is poikilothermy and homeothermy? How do these terms relate to those in the question above? (86)
10. List some advantages and disadvantages of endothermy. (86-87)
11. Distinguish between acclimation and acclimatization. (82-83)
12. What is the active temperature range? Preferred temperature range? Zone of thermal tolerance? Zone of thermal resistance? (86-88)
13. What is heliothermism? How do some heterothermic insects employ that to fly? (88)
14. How can amphibians regulate body temperature? Reptiles? (86-89)
15. What are some of the advantages and disadvantages of endothermy? (90-91)
16. What is countercurrent circulation and how does it function in heat regulation? (91, 93-94)
17. Why is it more difficult for endotherms or homeotherms to adjust to high temperatures than to low ones? (91-92)
18. How do desert mammals employ hyperthermia to maintain heat balance? (91-92)
19. Distinguish between torpor and hibernation. What are the metabolic characteristics of a hibernating mammal? (94-95)
20. How do bears differ from other hibernators? (94-95)
21. Why is a bear able to successfully starve itself? (94-95)
22. In what ways does temperature limit the distribution of plants and animals? (95-97).

Key Terms and Phases

convection	endothermy	countercurrent circulation
conduction	ectothermy	rete
radiation	active temperature range	critical temperature
thermal balance	preferred temperature range	hyperthermia
ambient temperature	incipient lethal temperature	torpor
core temperature	thermal tolerance	diapause
cold resistance	thermal resistance	estivation
chilling	acclimation	hibernation
hardening	acclimatization	evaporation
supercooling	heliothermism	
homeotherms	critical thermal maximum	
poikilotherms	proportional control	
heterotherms	thermal neutral zone	

Key Term Review

1. Heat transferred when a fluid circulates is _____.

2. _____ is the direct transfer of heat from one substance to another.

3. Heat is emitted from surfaces by _____.

4. _____ _____ is the temperature of the environment.

5. Fish can adjust to a seasonal change in water temperature by _____.

6. _____ is the metabolism of brown fat to increase body heat.

7. Frogs and salamanders are classified as _____ because their body temperature remains in equilibrium with the environment.

8. A dolphin prevents excessive heat lost from its fins by means of _____ _____.

9. A _____ is a net of intermingling blood vessels.

10. Arctic marine fish experience _____ which involves special chemicals in their body that prevent freezing.

11. An insect warming up in the sun employs _____.

12. _____ is a resting stage of no growth in insects.

13. _____ gain heat mostly from the environment.

14. _____ maintain a fairly constant internal body temperature.

15. _____ results when heat accumulates in a homeotherm raising its normal core temperature.

16. The temperature at which body insulation is no longer effective and body heat must be maintained by increased metabolism is the _____ _____.

17. Range of body temperatures within which a ectotherm carries out its daily activities is its _____ _____ _____.

18. Many woody plants convert sensitive cells into cold resistant ones by _____ _____.

19. To reduce energy demands during the heat of a summer day, bats reduce their metabolic rate to enter a condition known as _____.

20. During the winter bats congregate in caves, reduce their body metabolism, and reduce body temperature to ambient for a long period. This condition is called _____.

Self Test

True and False

1. _____ Poikilotherms at times can be functional endotherms.

2. _____ Bees and dragonflies can fly immediately in a cool morning without warming.

3. _____ Hibernating mammals accumulate metabolic CO_2 in their blood.

4. _____ A plant's ability to resist low temperature stress without injury is frost tolerance.

5. _____ When a mammal's body temperature drops rapidly below normal this is hyperthermia.

6. _____ Heat gained by an organism must equal heat lost.

7. _____ Large animals lose proportionally more heat to the environment than small ones.

8. _____ Hibernation is a dormancy experienced by some desert animals during the summer.

9. _____ The amount of heat that moves in or out of a body is proportional to the body's weight.

10. _____ Conduction is the movement of heat between a warm solid object and a cool one.

11. _____ In countercurrent circulation warm arterial blood exchanges heat with cool venous blood.

12. _____ The range of temperatures in which aquatic poikilotherms have their highest survival is thermal tolerance.

Matching

Associate each of the following physiological characteristics with poikilothermy (ectotherms) and homeotherms (endotherms) by placing a P or an H in front of the statement.

1. _____ High thermal conductance between body and environment.

2. _____ Burn energy rapidly.

3. _____ Energy production is largely anaerobic.

4. _____ Metabolic rate decreases as body size increases.

36

5. _____ Allocate more energy to biomass production than to metabolism.

Short Answer Questions

1. What special adaptations do desert plants employ to tolerate high temperatures?

2. If ectotherms lack physiological controls of their metabolic rate, how can they regulate their body temperature?

3. Why are terrestrial animals subject to more radical changes in their thermal environment than aquatic organisms?

4. When a plant dies from a rapid drop in temperature, what probably happened to the plant at the cellular level?

5. What characteristics of a plant, particularly the leaf, influence the amount of heat a plant absorbs? List two and explain how they function.

6. Why are small endotherms so rare?

ANSWERS

Key Term Review

1. convection
2. conduction
3. radiation
4. ambient temperature
5. acclimatization
6. thermogenesis
7. poikilotherms
8. countercurrent circulation
9. rete
10 supercooling

11. heliothermism
12. diapause
13. ectotherms
14. homeotherms
15. hyperthermia
16. critical temperature
17. active temperature range (ACT)
18. frost hardening
19. torpor
20. hibernation

Self Test

True and False

1. T	7. F
2. F	8. F
3. T	9. F
4. T	10. T
5. F	11. T
6. T	12. T

Matching

1. P
2. P
3. P
4. H
5. H

Short Answer Questions

1. Adaptations include small leaves, hold leaves parallel to the sun's rays, carry on photosynthesis in stems, shut down normal protein synthesis, and acclimate to high temperatures.

2. Poikilotherms can regulate their body temperature by seeking shade to cool, by seeking sun to warm, and by increasing or decreasing their body surface area exposed to the sun (proportional control).

3. Environmental temperatures for terrestrial animals ranges widely from day to night and seasonally. Aquatic organisms live in an environment in which temperature does not fluctuate as rapidly. Aquatic poikilotherms do not maintain any appreciable difference between their body temperature and the environment.

4. Ice crystals formed inside the plant cells, expanding the cell walls. When the plant tissues thaw, the cell contents spill out.

5. Characteristics that influence the amount of heat a plant absorbs are:
    (a) Lobing in the leaf which increases the surface area of the leaf available for cooling.
    (b) Increased transpiration to cool leaf by evaporation.
    (c) Hold the upper leaf surface parallel to sun's rays.

6. Small endotherms are rare because there is a lower limit to the amount of surface area relative to body weight that can be exposed before heat loss to environment exceeds metabolic heat production.

CHAPTER 7
LIGHT AND BIOLOGICAL CYCLES

Chapter Outline

The Nature of Light
Plant Adaptation to Light Intensity
   PAR Radiation
   Ultraviolet Radiation
Photoperiodism
   Circadian Rhythms

The Biological Clock
Critical Daylength
Tidal and Lunar Cycles
Seasonality
Summary

Learning Objectives

After completing this chapter you should be able to:

- Define the nature of light in the environment.
- Identify the fate of solar radiation as it strikes Earth.
- List the similarities and differences between shade-tolerant and shade-intolerant species of plants.
- Compare and contrast adaptations of terrestrial and aquatic plants to light in their respective environments.
- Explain the ecological significance of ultraviolet radiation.
- Recognize the significance of circadian rhythms to the biological clock.
- Compare and contrast different models of biological clocks.
- State the adaptive values of circadian rhythms.
- Discuss the mechanisms involved in the response of plants and animals to seasonal changes in the length of day.
- Give examples of the role of critical daylengths in regulating the activities of plants and animals.
- List phenological responses of plants and animals to light, temperature, and moisture variations created by changes in altitude and latitude.

Summary

After reading the chapter and before continuing with the material below read the Chapter Summary on page 113. You will find related material on the following pages: adaptation, 30-31; solar radiation, 36-38; thermal energy exchange, 80-82; photosynthesis, 154-159; tides, 329-330; succession, 659-663.

Study Questions

1. What is photosynthetically active radiation? (100, 103)
2. In what order are wave lengths attenuated in water? In terrestrial vegetation?

Study Questions

1. What is photosynthetically active radiation? (100, 103)
2. In what order are wave lengths attenuated in water? In terrestrial vegetation? (101-102)
3. What is compensation intensity? (102)
4. What is the leaf area index? (102)
5. What are the basic differences between shade-tolerant and shade-intolerant plants? (103-104)
6. Name two tree species very tolerant to shade, two intolerants, two intermediate, two shade tolerant. (Such information is relevant to the study of succession.) (Table 7.1).
7. What is the relationship between the amount of ultraviolet radiation and atmospheric depletion of the ozone layer? (105)
8. What are some defenses against UV radiation evolved by plants? (105)
9. What is a circadian rhythm? (106)
10. What is the free running cycle? (106)
11. How does the endogenous circadian rhythm become entrained to the exogenous rhythms of light and dark? (107)
12. What is the biological clock? (106)
13. What is critical daylength? (106)
14. How does daylength or photoperiodism influence the reproductive cycle in birds? In mammals? (109)
15. What is a short-day response? A long-day response? (108)
16. What is phenology? (110)
17. What is the relationship of seasonality to photoperiod in plants and animals? How is it influenced by climatic gradients? (110)
18. What is the relationship between tidal and lunar cycles to life of the intertidal zone? (110)

Key Terms and Phrases

extinction coefficient
compensation intensity
shade tolerance
photosynthetically active radiation
photoperiodism
circadian rhythms
entrainment

free-running cycle
Zeitgerber
biological clock
ultraviolet radiation
short-day organism
long-day organism
lunar cycle

melantonin
critical day length
seasonality
phenology
leaf area index
tidal cycle

Key Term Review

1. Light in the wavelengths between 400 and 740 nannometers is used in photosynthesis. These wavelengths are known as _____ _____ _____.

2. The point at which the amount of light penetrating lake water is too little for photosynthesis is the _____ _____.

3. The light intensity at which plants can no longer carry out sufficient photosynthesis to meet their energy requirement is the _____ _____.

4. Plants that can carry on photosynthesis at low light intensity are _____.

5. Innate rhythms of activity and inactivity that approximate 24 hours of time are called _____ _____.

6. The period of innate rhythms named in the above question is called the _____ _____.

7. _____ are environmental stimuli that set biological clocks.

8. _____ is a process that is analogous to setting a watch to the correct time.

9. _____ _____ defines a period of light that when reached either inhibits or promotes a response.

10. Many biological events recur with the changing of seasons. The study of the causes of the timing of these events is called _____.

11. Laboratory studies suggest that the salt marsh periwinkles show entrainment of activity cycles related to _____ _____.

12. Most of the _____ _____ that reaches Earth is filtered out by the ozone layer.

13. _____ is a hormone produced more during the dark cycle of a circadian rhythm. It is involved in an animals seasonal reproductive cycle.

14. The timekeeper of physical and physiological activity of living things is the _____ _____.

15. Organisms whose reproduction and other activities are stimulated by daylengths shorter than the critical daylength are know as _____ organisms.

16. Biological events recurring with passage of seasons describe _____.

17. The total surface area of leaves above a given area of ground is know as _____ _____ _____.

18. Recurring daily and seasonal changes in the activity of organisms in response to daily seasonal daylength is know as _____.

Self Test

True and False

1. _____ Phenology is the study of the distribution of plants and animals as determined by their responses to light.

2. _____ The circadian rhythm has a periodicity of approximately 24 hours.

3. _____ The most important synchronizer for circadian rhythms is temperature.

4. _____ The critical daylength for most plants and animals falls somewhere between 10 and 14 hours.

5. _____ Most of the sunlight that strikes a forest is intercepted by the forest floor.

6. _____ Diapause is a process generally associated with mammals over winter.

7. _____ Shade tolerant plants have a higher rate of leaf respiration than shade-intolerant plants.

8. _____ You would expect to find green algae in deep water.

9. _____ Blue wave lengths are easily absorbed by water.

10. _____ Shade-intolerant plants have less chlorophyll per unit of leaf weight than shade-tolerant plants.

11. _____ Circadian rhythms are genetic.

Matching.

Indicate whether the following are long-day or short-day responses in the organisms indicated by placing an **S** for short-day and **L** for long-day

1. _____ Antler growth in white-tailed deer.

2. _____ Reproductive activities in doe and buck white-tailed deer.

3. _____ Food storage by flying squirrels.

4. _____ Acquisition of a white coat in snowshoe hares in late fall and early winter.

5. _____ Spring breeding behavior in birds.

6. _____ Goldenrod's bloom in the fall.

7. _____ Woodland trilliums bloom in early spring.

Match the lettered response with the numbered questions.

A. Entrainment   B. Zeitgerber   C. Free-running   D. Master clock

8. _____ A process that is similar to setting a watch to correct time.

9. _____ An environmental stimulus such as the amount of daylight that sets a biological clock.

10. _____ Reset by light, it then resets other biological clocks in an organism.

11. _____ The period of a circadian rhythm.

Short Answer Questions

These diagrams apply to short-day plants.

1. Is this plant in flower or not? Why?

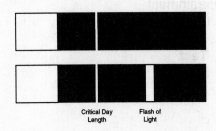

Critical Day Length   Flash of Light

2. Assume light is flashed as indicated in (b). What effect will it have on the plant?

3. List three characteristics of circadian rhythms.

4. What are the ecological and thus survival advantages of photoperiodicy in plants and animals?

5. What are the two periodicities or rhythms that influence the lives of plants and animals? How are they synchronized?

6. Many garden catalogs categorize perennial plants according to their ability to grow in sun, partial shade, or shade. What problems do gardeners face if they unknowingly planted a sun-loving or shade-intolerant plant in the shade of trees?

ANSWERS

Key Term Review

1. photosynthetically active radiation
2. extinction coefficient
3. compensation intensity
4. shade tolerant
5. circadian rhythm
6. free running cycle
7. zietgerbers
8. entrainment
9. critical daylength
10. phenology
11. tidal cycles
12. ultraviolet light
13. melantonin
14. biological clock
15. short-day
16. seasonality
17. leaf area index
18. photoperiodicity

Self Test

True False

1. F
2. T
3. F
4. T
5. F
6. F
7. F
8. F
9. F
10. T.
11. T

Matching

1. L
2. S
3. S
4. S
5. L
6. S
7. L
8. A
9. B
10. D
11. C

Short Answer Questions

1. The short-day plant is in flower because the dark cycle exceeds the critical daylength of this plant. The plant responds to short-day conditions and flowers.

2. The flash of light in the dark cycle will cause the plant to respond to long-day conditions and fail to flower.

3. Three characteristics include a) free-ruuning until entrained to environmental light cycle; b) inherited (genetic), not learned; c) relatively temperature insensitive.

4. Photoperiodicity allows organisms to keep their activities in synchrony with daily and seasonal environmental changes. It is especially important in synchronizing reproductive activities with favorable seasonal environment.

5. The two periodicities are the endogenous circadian rhythms and the exogenous environmental cycle of light and dark. Light is the key timesetter that sets the biological clock, whose mechanism is the light and dark sensitive periods of the circadian rhythm.

6. The sun plant will grow rapidly in height "seeking" light. Such growth results in weak-stemmed plants with thin leaves. The plant will fail to flower.

CHAPTER 8
NUTRIENTS

Chapter Outline

Essential Nutrients
Nutrient Sources and Cycling
Nutrients and Plants
    Calcicoles and Calcifuges
    Plants of Serpentine and Toxic Soils

Plants of Saline Habitats
Atmospheric Pollution Injury
Nutrients and Consumers
Summary

Learning Objectives

After completing this chapter you should be able to:
- Distinguish between micronutrients and macronutrients.
- List the major macronutrients and the biological importance of each.
- List the major micronutrients and the biological importance of each.
- Summarize how plants and animals meet their nutrient requirements.
- Tell why different levels of nutrients influence competitive abilities among plants.
- Compare and contrast terrestrial and aquatic nutrient cycling.
- Discuss a basic model of nutrient cycling.
- Define calcicole and calcifuge.
- List adaptations of plants growing on calcareous, acid, serpentine, and saline soils.
- Define ecotypes and endemics.
- Discuss the various types of atmospheric pollutants that affect the growth of plants.
- Describe the nutrient requirements of animals.
- Distinguish between nutrient quantity and nutrient quality.

Summary

After reading through this chapter and before continuing with the material below, read the Chapter Summary on pages 127-128. You will find related material on the following pages: adaptation, 30-31; plant adaptation to drought, 72; soil chemistry, 130-134; decomposition 159-165; ruminants, 183; carbon cycle, 199-206; nitrogen cycle, 206-210; sulfur cycle, 210-212; phosphorus cycle, 212-213; heavy metals, 213-214; acid deposition 214-216.

Study Questions

1. Distinguish between a macronutrient and a micronutrient. What are some examples of each? (116)
2. Why does fertilization or nutrient enrichment reduce plant diversity? (120-121)
3. Distinguish between a calcicole and a calcifuge? What are the outstanding characteristics of each? (121-122)

4. What is a serpentine soil?   (122)
5. What is the effect of heavy metals on plant tolerance and distribution? (122)
6. What are halophytes? Where would you find them? (123-124)
7. Comment on the response of vegetation to three levels of air pollution: low, intermediate, and high. (124)
8. How does pollution stress affect forest trees? (124)
9. How does the nutrient level in plants affect the well-being of animals, especially herbivores? (124-127)
10. What is a nutrient budget of an ecosystem? (118-120)
11. What are some of the inputs and outputs in the nutrient budget of a forest? (116-119)
12. What is the importance of wetfall, dryfall, throughfall, and stemflow? (116-119)
13. Describe the short-term cycling of nutrients in a forest. (118-119)
14. Contrast nutrient cycling in aquatic ecosystems with that of terrestrial systems. (119-120)

Key Terms and Phrases

micronutrients	stemflow	halophyte
macronutrients	calcicole	endemic
dryfall	calcifuge	ecotype
wetfall	neutrophile	ruminant
throughfall	serpentine	chlorosis
leaching		

Key Term Review

1. Nutrients that are needed by plants and animals in relatively large amounts are known as _____.

2. Nutrients carried to plants by precipitation is called _____.

3. Over one-half of the input of calcium and nitrates is from airborne particles and aerosols , called _____.

4. True _____ grow on soils with a high pH.

5. _____ soils are high in nickel and iron.

6. Plants tolerant of high concentration of salts are called _____.

7. Plants adapted to a set of local conditions are known as _____.

8. _____ are associated with acid soils.

9. Approximately five percent of total rainfall reaching a forest floor is _____.

10. Plants that grow in both acidic and mildly alkaline soils are _____.

11. Species restricted to certain specialized habitats are _____.

12. Large grazing herbivores with four-chambered stomachs are called
    _____.

13. Loss of water-soluble nutrients from leaves and other organic material is called
    _____.

14. Acid-requiring plants planted in high lime soils exhibit a yellowing of leaves called
    _____.

Self Tests

True and False

1. _____ Macronutrients are chemicals needed in large quantities by plants and animals.

2. _____ Iron is a micronutrient.

3. _____ Carbon, oxygen, nitrogen, and calcium are all macronutrients.

4. _____ Serpentine soils are high in calcium and low in magnesium.

5. _____ Sodium availability can be an important influence in deciding the distribution of herbivorous mammals.

6. _____ Atmospheric pollutants, especially CO_2, are damaging to vegetation.

7. _____ Halophytes are adapted to saline conditions.

8. _____ Ruminants face severe mineral deficiencies in spring.

9. _____ Trace elements are called macronutrients.

10. _____ Carnivores rarely have nutrient deficiencies because they consume other animals.

11. _____ An aquatic ecosystem depends on an input of nutrients from the lands surrounding them.

12. _____ A major atmospheric pollutant damaging to vegetation is SO_2.

Matching

Match the lettered nutrient with the numbered function of that nutrient:

A. Phosphorus   F. Iodine   K. Manganese
B. Sulfur   G. Aluminum   L. Iron
C. Nitrogen   H. Cobalt   M. Oxygen
D. Carbon   I. Calcium   N. Sodium
E. Boron   J. Magnesium   O. Copper

1. _____ This nutrient is a building block of protein.

2. _____ Nutrient that plays a major role in energy transfer.

3. _____ Nutrient that shows some correlation with the distribution of grazing herbivores.

4. _____ This nutrient reaches toxic levels in acid soils.

5. _____ The building block of all organic molecules.

6. _____ Needed in large quantities by plants, this nutrient is involved in maintaining turgor and regulating the movement of the stomata.

7. _____ Essential for biological oxidation.

8. _____ Important in thyroid metabolism.

9. _____ Ruminants use this nutrient to synthesize Vitamin B_{12}.

10. _____ Deficiency in ruminants causes a serious disease, grass tetany.

11. _____ Deficiency of this nutrient results in anemia.

12. _____ In plants it enhances the transfer of electrons from water to chlorophyll and is important in the synthesis of fatty acids.

13. _____ Its deficiency in plants causes stunted grow in roots and leaves and yellowing leaves.

14. _____ A basic constituent of protein.

15. _____ Found in chloroplasts and used to regulate the photosynthetic rate.

Associate each of the following characteristics with a (**A**) calcicole plant, (**B**) calcifuge plant, or (**C**) a serpentine soil plant.

16. _____ Grows best in soils of high pH.

17. _____ Tolerant of aluminum ions.

18. _____ Grows in the presence of high concentrations of nickel and zinc.

19. _____ Suffers from chlorosis in the presence of calcium.

20. _____ Many are endemic.

Short-Answer Questions

1. What governs the supply of nutrients to living organisms?

2. Certain species of plants have become adapted to soils made toxic by heavy metals. How might these adaptations come about?

3. Contrast short-term nutrient cycling and long-term nutrient cycling in a forest ecosystem.

4. Describe nutrient cycling in aquatic ecosystems, emphasizing the difference with terrestrial ecosystems.

ANSWERS

Key Term Review

1. macronutrients
2. wetfall
3. dryfall
4. calcifuge
5. serpentine
6. halophytes
7. ecotypes

8. calcicoles
9. stemflow
10. neutophiles
11. endemic
12. ruminant
13. leaching
14. chlorosis

Self Test

True and False

1. T
2. T
3. T
4. F
5. T
6. F
7. T
8. T
9. T
10. F
11. T
12. T

Matching

1. C
2. A
3. N
4. G
5. D
6. I
7. M
8. F
9. H
10. J

11. L
12. K
13. E
14. B
15. O
16. A
17. B
18. C
19. B
20. C

Short Answer Questions

1. The supply of available nutrients is governed by a) inputs into the detrital pool by producers and consumers; b) rate of decomposition and nutrient release from organic material; c) amounts of detritus and nutrient going into long-term storage; c) release of nutrients from long-term reserve.

2. Plants growing on toxic soils evolved ecotypes tolerant of heavy metals, but are poor competitors in normal habitats. This evolution probably involved selection of tolerant seedling from plants of surrounding areas, continued selection for metal tolerance, and selection to survive in a harsh environment.

3. Short-term cycling of nutrients in terrestrial ecosystem involves the decomposition of leaves and dead herbaceous plants; the uptake of these nutrient by plant roots; and the return of these nutrients by litterfall, throughfall, and stemflow. Long-term cycling involves storage of nutrients in limbs, trunk, bark, and root, and their death and slow decomposition. This slow decomposition gradually releases nutrients to the soil where they are again available to plants.

4. Nutrient cycling in aquatic ecosystems involves inputs from water draining from the surrounding land, sediments, leaves and other organic matter deposited in the water, and from precipitation. Nutrient cycling is mostly short term involving turnover of nutrients in phytoplankton and zooplankton. Unlike terrestrial ecosystems, aquatic ecosystems lack any long-term biological retention.

CHAPTER 9
SOILS

Learning Objectives

After completing this chapter you should be able to:
- Define soil.
- Explain why soil is a living system.
- Describe how soil is formed.
- Identify soil horizons and describe the characteristics of each.
- Explain how properties of color, texture, structure, and moisture content vary in different soils.
- Define soil texture.
- Distinguish between sand, silt, and clay.
- Discuss why soil moisture is an important characteristic of soil.
- Summarize the factors that influence the chemistry of soil.
- Outline the specific characteristics of the organic horizon and the major biological cycles that occur in it.
- Define podzolization, laterization, calcification, salinization, and gleyization.
- Explain how each of the above processes is involved in soil development.
- Explain how soil develops through time.
- List the major soil types.
- Discuss soil mapping and the meaning of catenas, toposequences, soil types , and soil series.

Summary

After reading through the chapter and before continuing with the material below, read the Chapter Summary on pages 147-148. You will find related material on the following pages: local water cycle, 67; soil-plant-atmosphere continuum, 70-71; nutrient sources and cycling, 116-120; calcicoles and calcifuges, 121-122; nutrients and consumers, 124-127; decomposition and mineralization, 159-165.

Study Questions

1. What is soil? (130)
2. What is a soil profile? (130)
3. What are soil horizons? Name them and give major characteristics of each. (130-131)
4. What physical properties distinguish different soils? (131-133)
5. What is meant by soil structure? What are peds? (132)
6. What determines the texture of soil? (132)
7. Distinguish between field capacity, available water capacity, and permanent wilting point. (133)
8. What is the unique structure of clays? What elements make up the basic clay mineral? 134)
9. What are micelles? (134)
10. What are exchangeable ions and what is cation exchange capacity? (135)
11. What five independent factors are involved in soil formation? (137-138)
12. What are the major types of parent material? (138)
13. What is involved in the weathering of soil? (138)
14. How do living organisms influence soil development? (136-139)
15. What is humus? How is it formed? (139-140)
16. Distinguish between mull, mor, and moder types of humus. (139-140)
17. What is the difference between the soils of coniferous forest, deciduous forests, tropical forests, grasslands, and tundra? (141)
18. What is podzolization, calcification, salinization, laterization, and gleization? (140-145)
19. What are the eleven major soil orders? (143)
20. Distinguish between a soil series, a catena, a toposequence, a chronosequence, and a soil association in soil mapping and classification. (144-146)
21. What is soil erosion? Distinguish between surface, rill, and gully erosion. (147)

Key Terms and Phrases

soil	gley	spodosols
soil profile	gleization	alfisols
soil horizons	alluvial deposits	utisols
soil texture	lacustrine deposits	oxisols
silt	marine deposits	mollisols
sand	residual materials	caliche
clay	regolith	salinization
soil structure	decomposition	toposequence
aggregrates	mineralization	catena
field capacity	humus	chronosequence

58

available water capacity   organic horizon   peds
permanent wilting point   mull   soil series
micelles   mor   soil association
cation   moder   soil complexes
cation exchange capacity   podzolization   till
percent base saturation   laterization   weathering
loess   calcification   leaching
aridosols

Key Term Review

1. A vertical cut through a body of soil exposes the _____ _____.

2. The _____ of a soil is determined by the types and sizes of soil particles in it. Particles are classified as _____, _____, and _____.

3. _____ consists of particles less than 0.0002 mm.

4. Soil particles are held together in clusters called _____ or _____.

5. The maximum amount of water a soil will hold is the _____ _____.

6. _____ are plate like particles is a soil particle of clay and humus carrying electrical charges on the surface.

7. The number of negatively charged sites on a soil article that can attract a positively charged _____ is called the _____ _____ _____.

8. Soil material carried from one place to another by wind is known as _____.

9. _____ is partially decomposed organic matter.

10. The biological and chemical breakdown of the _____ into soil is called _____.

11. _____ removes soluble nutrients from the soil and organic matter.

12. The _____ _____ _____ is the supply of water available to plants growing on well-drained soil.

13. The _____ _____ _____ occurs when the soil is dry and plants cannot extract enough water to meet their demands.

14. Soils that have a B horizon rich in iron and aluminum formed by _____.

15. In the tropics heavy rainfall and high temperatures develop an unique soil type called _____ by a soil forming process called _____.

16. In poorly drained soils _____ can result in a soil with a compacted horizon characterized by ferrous oxides.

17. Percent of sites on micelles that are occupied by ions other than hydrogen is called _____ _____ _____.

18. Soil material transported by glacial ice is _____.

19. The breakdown of organic material into humus is accomplished by _____ and _____.

20. Two types of humus formation in temperate forest soil are _____ and _____.

21. _____ is characterized by a layer of compacted organic matter distinct from the mineral soil.

22. The insect mull of P. E. Muller is _____.

23. _____ is a soil forming process that involves the accumulation of calcium in lower horizons. Soils formed by this process are mostly _____.

24. A group of related soils is known as a soil _____.

25. A sequence of related soils differing in profile development because of differences in ages make up a _____.

26. The basic unit of soil classification is the soil _____.

27. Desert soils characterized by little organic matter and a high base content are _____.

28. Developmental layers in the soil are _____ _____.

29. Alkaline, rock-like salt deposits on the surface of desert soils are called _____.

30. Reddish soils resulting from the release of iron from silicates common in southeastern United States are _____.

Self Test

True and False

1. _____ Soil materials transported by water from one place to another form alluvial deposits.

2. _____ Podzol soils may develop in temperate forests.

3. _____ Soils with high water content will be deficient in oxygen.

4. _____ The texture of soil is determined by the proportion of sand, silt, and clay in it.

5. _____ A cross-section of a soil in an area is called a horizon.

6. _____ The soils in tropical rain forests are classified as aridosols.

7. _____ Micelles are plate-like clay particles in the soil.

8. _____ The number of positively-charged sites on a soil particle that can attract negatively-charged is called the cation exchange capacity.

9. _____ Reddish color of a soil horizon result from the presence of oxidized iron.

10. _____ Clay particles control many important properties of soil.

11. _____ Groups of soils developed from the same soil material are called chronosequences.

12. _____ All organic matter is converted to true humic substances in mull humus.

13. _____ Soils are divided into recognizable units called soil series.

14. _____ Soil forming process that involves downward leaching and partial removal of soluble salts and colloidal materials is called calcification.

15. _____ Clay content of a soil has an important bearing on its cation exchange capacity.

16. _____ Soil texture influences the movement and retention of water in the soil.

17. _____ Decomposing organic matter forms an unincorporated mat on top of mineral soil in mull humus.

18. _____ Hydrogen ions are strongly attracted to micelles.

19. _____ Soil acidity has little effect on nutrient availability.

20. _____ Topographic position has little influence on soil moisture.

Matching

Associate the horizon descriptions with the appropriate horizon labels:

Horizon Labels: **O, A, B E, C, R**

1. _____ Zone of maximum biological activity.

2. _____ Zone of maximum leaching.

3. _____ Nonsoil horizon

4. _____ Clay accumulates here.

5. _____ Zone of active weathering

6. _____ Accumulation of organic matter and loss of clay.

Associate each of the following soil processes or description with the appropriate term:

A. Gleization; **B,** Laterization; **C,** Salinization; **D,** Podzolization; **E,** Calcification.

7. _____ Thick A horizon with a B horizon in which calcium accumulates.

8. _____ B horizon rich in iron; light colored A horizon; acid.

9. _____ Heavy rainfall and high temperatures result in uniformly weathered soil with silica leached out; iron and aluminum oxides are retained.

10. _____ High in base content, low in humus; typical of areas with low precipitation.

11. _____ Compact horizons rich in organic matter; poorly decomposed organic matter; typical of area with poor drainage.

Short Answer Questions

1. What causes soils to become acid?

2. Why is the clay-humus complex the key to the availability of soil nutrients?

3. In what way do living organisms influence the development of soil?

ANSWERS

Key Term Review

1. soil profile
2. structure; sand, silt, clay
3. clay
4. peds or aggregrates
5. field capacity
6. micelles
7. cations, cation exchange capacity
8. loess
9. humus
10. regolith, weathering
11. leaching
12. available water capacity
13. permanent wilting point
14. podzolization
15. oxisols, laterization
16. gleization
17. percent base saturation
18. till
19. decomposition, mineralization
20. mor, mull
21. mor
22. moder
23. calcification, mollisols
24. catena
25. chronosequence
26. series
27. aridosols
28. soil horizons
29. caliche
30. utisols

True and False

1. T	11. F
2. T	12. T
3. T	13. T
4. T	14. F
5. F	15. T
6. F	16. T
7. T	17. F
8. T	18. T
9. T	19. F
10. T	20. F

Matching

1. O
2. E
3. R
4. B
5. C
6. A
7. E
8. D
9. C
10. B
11. A

Short Answer Questions

1. Soil acidity results from the removal of bases by the leaching effects of water, withdrawal of exchangeable ions by plants, release of organic acids by roots and microorganisms, and the dissociation of $CaCO_3$.

2. The clay humus complex is the site of cation exchange between soil and soil water or solution. The availability depends upon how closely held the cations are held on the clay and humus particles. Negative charges enable the soil to prevent leaching of its positively charged nutrient cations. Cations in the soil solution are constantly being replaced by exchanges with cations on clay and humus particles.

3. Plants, animals and microorganism have a pronounced influence on soil development. Plant roots break up the substrate and add organic matter to the upper layer of soil and withdraw nutrients from deep layers and bring them to the surface. Action of bacteria, and fungi reduce organic matter to humus. Earthworms and other invertebrates mix organic matter in the mineral soil, while burrowing animals mix material from lower layers with the upper.

PART 3

THE ECOSYSTEM

CHAPTER 10
CONCEPT OF THE ECOSYSTEM

Chapter Outline

Components of Ecological Systems
Essential Process
Photosynthesis
Decomposition
Summary

Learning Objectives

After completing this chapter you should be able to:
- Describe the basic components of ecosystems.
- Explain the major reactions in photosynthesis.
- Distinguish C_3, C_4, and CAM types of photosynthesis.
- Explain the decomposition process and the type of decomposer organisms involved.
- Distinguish between immobilization and mineralization.
- Discuss the influences on decomposition rates.
- Contrast the decomposition process in aquatic ecosystems vs. terrestrial systems.

Summary

After reading this chapter and before proceeding with the material below, read the Chapter Summary on pages 165-166. You will find related material on following pages: characteristics of water, 64; soil humus, 139-40; primary production, 169-175; secondary production, 178-181; detrital food chain, 185-187; energy flow, 189-191; trophic levels, 191-192; food webs, 192-193, 336, 621-623; biogeochemical cycles, 196-221.

Study Questions

1. Define biosphere, hydrosphere, lithosphere. (152)
2. What is meant by an ecosystem? By whom and when was the term coined? (152-153)
3. What are the three major components of an ecosystem and how are they related? (153)
4. What is the driving force of an ecosystem? (153)
5. What are the two essential functional processes in an ecosystem? Why? (154)
6. What is photosynthesis? Explain how the sun's energy is used? (154-155)
7. What are C_3, C_4, and CAM plants? What are the differences among them? Where are they most likely to be found? (154-157)
8. What is the ecological significance of C_3 and C_4 methods of photosynthesis? (154-157)

9. What is LAI? What is its significance? (158)
10. What is decomposition? Why is it important ecologically? (159)
11. Name the major groups of decomposers and describe their functions. (159-163)
12. What are the sequential stages of decomposition? (161)
13. Distinguish between nutrient immobilization and nutrient mineralization. (162-163)
14. What is the importance of detritivores in decomposition? (163)
15. Describe the difference between the decomposition process of plant and animal matter? (163)
16. What influences the rate of decomposition? (163-164)
17. Contrast aquatic decomposition with terrestrial decomposition. (164-165)

Key Terms and Phrases

biosphere
lithosphere
hydrosphere
ecosystem
producers
autotrophs
consumers
heterotrophs
detritus
photosynthesis

photorespiration
C_3 plants
C_4 plants
CAM plants
leaf area index
decomposition
microflora
obligate anaerobes
fermentation

detritivores
microbivores
leaching
fragmentation
catabolism
mineralization
anabolism
rhizosphere
rhizoplane

Review of Key Terms

1. _____ _____ is the ratio of surface area of leaves to area on the ground available to the plant.

2. The _____ is the zone including Earth's surface and surface water, the adjacent atmosphere, and that part of underlying crust in which life can exist.

3. The liquid part of the biosphere is the _____.

4. _____ can convert simple inorganic compounds, including carbon dioxide, to into complex organic molecules. They are also called _____.

5. The solid part of the biosphere is the _____.

6. Conversion of light energy from the sun by plants and phytobacteria into chemical energy used to produce carbohydrates is _____.

7. _____ is respiration that occurs in plants in the light.

8. The breakdown of dead organic matter into simple substances by organisms of decay is _____.

9. Aiding in the breakdown of organic matter are fungi and algae, collectively called _____.

10. Feeding on dead organic matter are the _____.

11. Preying on fungi and bacteria are the _____.

12. _____ is the washing of soluble substances from dead organic matter and upper layers of soil.

13. _____ is the breakdown of complex molecules to simpler ones, whereas _____ is the metabolic synthesis of complex substances from simpler ones.

14. _____ require a supply of organic matter for food. They are also called _____.

15. _____ _____ are microbes that cannot live in the presence of oxygen.

16. The reduction leaves and other organic matter into smaller pieces is _____.

17. The breakdown of glucose and other substances under anaerobic conditions is _____.

18. Breakdown of humus and other substances in the soil into inorganic substances is _____. Some of these substances may be converted into microbial or plant tissue; making it unavailable to other organisms in a process called _____.

19. The surface of the root is called the _____ and the soil about the root is the _____.

Self-Test

True and False

1. _____ C$_4$ plants are most likely to be found in warm climates where light intensities are high.

2. _____ In aquatic ecosystems, especially lakes and oceans, bacteria concentrate nutrients rather than release them.

3. _____ An autotroph depends upon energy fixed elsewhere for its energy source.

4. _____ The major source of dissolved organic matter in lakes is phytoplankton and zooplankton.

5. _____ Decomposition of animal material is more complex than decomposition of plant material.

6. _____ Facultative anaerobic bacteria cannot live in the presence of oxygen.

Matching

Three major types of plants based on their methods of photosynthesis are C_3, C_4, and CAM. Associate each of the characteristics given below with the appropriate type of photosynthesis. Use **A** for C_3, **B** for C_4, and **C** for CAM.

1. _____ Can carry on photosynthesis at relatively high temperatures.

2. _____ Stomata closed by day, open by night.

3. _____ Can carry on photosynthesis at low internal CO_2 levels.

4. _____ Less efficient at photosynthesis.

5. _____ Fixes CO_2 as malic and aspartic acid.

6. _____ Characteristic of warm weather grasses.

7. _____ Found in desert cacti.

8. _____ Photosynthetic efficiency reduced by photorespiration.

9. _____ Rarely become light saturated.

10. _____ Fixes CO_2 by night.

11. _____ Vascular bundles surrounded by a layer of dark cells, rich in chlorophyll.

Short Answer Questions

1. Straw and dead grass were plowed into the soil where they were attacked by fungal and bacterial decomposers. The next spring the new plant growth showed signs of nitrogen deficiency.
a. If decomposers are supposed to recycle nutrients, why were the plants experiencing a shortage of nitrogen.

b. By what process will inorganic nitrogen eventually be returned to the soil?

2. Leaf litter enclosed in fine-meshed bags and placed on the forest floor did not decompose as rapidly as the open leaf litter around them. Why did the bagged leaf litter decompose more slowly?

3. What is the major role of each of the following in the decomposition process:

A. Detritivores:

B. Microbial grazers:

C. Bacteria and fungi:

D. Bacteria in aquatic ecosystem:

E. Zooplankton in aquatic ecosystems:

Labeling

Assume that the dashed line in the diagram below represents the boundary of an ecosystem. Within the boundary are three major subsystems. Arrows indicate the relationships and "forcing functions" between the subsystems. Identify each of the subsystems and functions by placing an appropriate name or term in the spaces provided.

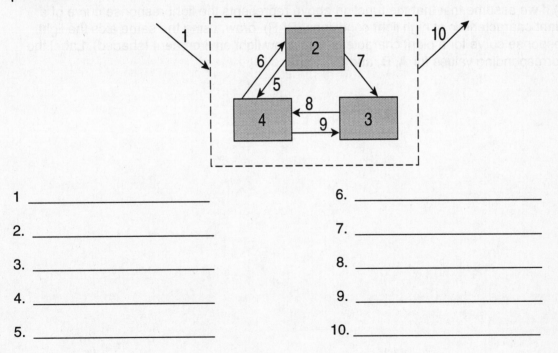

1 _____   6. _____

2. _____   7. _____

3. _____   8. _____

4. _____   9. _____

5. _____   10. _____

The figure below is a typical light response curve describing the relationship between photosynthesis (Pn) and photosynthetically active radiation (PAR) Define the three points in the figure labeled as **A**, **B**, and **C**.

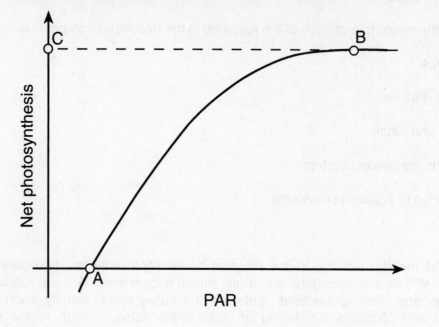

b) If we assume that that the function above represents the light response curve of a plant characteristic of high light conditions (PAR), draw, using the same axis the light response curve for a plant characteristic of a low light environment (shaded). Label the corresponding values for **A**, **B**, and **C**.

ANSWERS

Key Term Review

1. leaf-area index
2. biosphere
3. hydrosphere
4. autotrophs, producers
5. lithosphere
6. photosynthesis
7. photorespiration

8. decomposition
9. microflora
10. detritivores
11. microbivores
12. leaching
13. catabolism, anabolism

14. heterotrophs, consumers
15. anaerobic bacteria
16. decomposition
17. fermentation
18. mineralization, immobilization
19. rhizosphere, rhizoplane

Self Test

True and False

1. T
2. T
3. F
4. T
5. F
6. F

Matching

1. B
2. C
3. B
4. A
5. B/C
6. B

7. C
8. A
9. B
10. C
11. B

Short Answer Questions:

1. **a.** The straw and dead grass have a high carbon to nitrogen ratio. The supply of nutrients relative to energy source for microorganisms is low, reducing the activity of microorganisms. As long as the supply of energy is high and the demand for nitrogen is greater than available, microorganisms will incorporate or immobilize it in their biomes, making nitrogen unavailable for plants. Only when the CNN ratio declines and nitrogen is no longer in demand will nitrogen become available to plants. The solution is to reduce the CNN it by adding a form of easily available nitrogen to the field.

   b. Inorganic nitrogen will be returned to the soil by mineralization.

2. Leaf litter in the bag decomposed more slowly because the larger detritivores were excluded and fragmentation of the leaves was greatly reduced.

3. A. Detritivores fragment litter.

   B. Microbial grazers feed on bacteria and fungi, stimulating microbial growth.

   C. Bacteria and fungi bring about mineralization of nutrients.

   D. Aquatic bacteria concentrate nutrients in the water into bacterial biomass.

   E. Zooplankton release nutrients to the water.

Labeling

1. Input
2. Producers
3. Consumers
4. Organic matter, nutrients
5. Litterfall
6. Uptake, translocation
7. Consumption
8. Deposition
9. Decomposition
10. Output

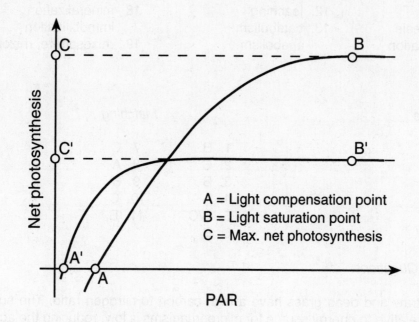

A = Light compensation point
B = Light saturation point
C = Max. net photosynthesis

CHAPTER 11
ECOSYSTEM ENERGETICS

Chapter Outline

The Nature of Energy
    Energy Defined
    Laws of Thermodynamics
Primary Production
    Energy Allocation
    Biomass Distribution
    Ecosystem Productivity
    Estimating Primary Productivity

Secondary Production
Food Chains
    Components
    Major Food Chains
Models of Energy Flow
    Trophic Levels and Ecological
      Pyramids
Summary

Learning Objectives

After completing this chapter you should be able to:

- Define energy and distinguish between potential and kinetic energy.
- Explain the laws of thermodynamics.
- Distinguish between production and productivity and between gross primary production and net primary production.
- Discuss energy allocation in plants.
- Contrast the productivity of different types of ecosystems.
- Describe briefly the different methods of estimating primary production.
- Explain the nature of secondary production.
- Describe the major components of food chains and food webs.
- Distinguish between the various feeding groups.
- Explain how herbivores convert plant tissue to animal tissue.
- Distinguish between detrital and grazing food chains and how the two interact.
- Describe trophic levels and ecological pyramids.

Summary

After reading this chapter and before continuing with the material below, read the Chapter Summary on pages 193-194. You will find related material on the following pages: biotic influences on soil, 138-140; the living soil, 136-137; process of photosynthesis, 154-159; photosynthetic efficiency, 159; process of decomposition, 161-165; plant-herbivore interactions, 522-524; plant defenses, 524-530; predator-prey relations, 530-541; intraguild predation, 542-543.

Study Questions

1. Define energy and distinguish between potential and kinetic energy. (168)
2. What is necessary for energy flow? (168)
3. What are the first and second laws of thermodynamics? How do they relate to ecology? (169)
4. What is primary production? Primary productivity? Gross primary production? Net primary production? Respiration? Standing crop biomass? (169-170)
5. How do plants allocate net primary production? (170-171)
6. What is the ecological significance of root/shoot ratios to production in terrestrial ecosystems? (172).
7. What influences primary productivity? (172-173)
8. What ecosystems have high productivity? Low productivity? Why? (174-176; Figures 11.6, 11.7; Table 11.1.)
9. What are methods employed to estimate primary production? (175-178)
10. What is secondary production? (179)
11. By means of a formula describe components involved in secondary production. (179)
12. What is assimilation efficiency? Growth efficiency? (180-181)
13. What is the difference in energy allocation between homeotherms and poikilotherms? (181)
14. What is a food chain? A food web? (181)
15. Describe the various feeding groups. What is the functional role of each? (181-184)
16. What are the two major food chains? How are they interrelated? (185-191)
17. What is trophic level? Relate trophic levels to three kinds of ecological pyramids. (191-192)
18. Why does the number of links in a food chain rarely exceed four? (192)
19. What problems are there with the concept of trophic levels? (192)
20. Theoretically what happens if a species is removed from a food web? Is the effect different if the species removed is a generalist? A top predator? (193)

Key Terms and Phrases

energy	compensation intensity	carnivore
potential energy	biomass	omnivore
kinetic energy	light-dark bottle method	decomposer
first law of thermodynamics	chlorophyll concentration	parasite
second law of thermodynamics	carbon dioxide flux	parasitoid
endothermic	dimension analysis	scavenger
exothermic	heat increment	saprophyte
entropy	secondary production	root/shoot ratio
primary production	community production	harvest method
gross primary production	food chain	trophic level
net primary production	food web	pyramid of numbers
respiration	grazing food chain	pyramid of biomass
growth	detrital food chain	pyramid of energy
storage	standing crop biomass	ruminant
accumulation	herbivore	trophospecies
reserve formation	corprophagy	biophage

recycling   assimilation efifciency   saprophage

Key Term Review

1. The number or biomass of squirrels in a woodlot in September is their
   _____ _____ _____.

2. Food assimilated/food ingested is _____ _____.

3. Gross primary production - net primary production equals _____.

4. Dead organic matter is the base of the _____ _____ _____.

5. Energy capable of and available for work is _____ _____.

6. In energy transformation much of the energy is degraded as heat. Increase in
   disorder of energy is _____.

7. When heat is involved in an energy transaction, the reaction is _____.

8. Energy that is due to motion and results in work is _____ _____.

9. In any energy transfer or transformation, no gain or loss in total energy takes place.
   This fact is known as the _____ _____ ____ _____.

10. When energy from outside surroundings is paid into a system to raise it to a higher
    energy state, the reaction is _____.

11. The _____ ____ _____ states that when energy tends to go from a
    more organized and concentrated state to a less organized state.

12. _____ _____ is the total amount of energy fixed by
    photosynthetic and chemosynthetic autotrophs. it is better expressed as _____
    _____ _____.

13. The total amount of organic matter manufactured by producers and consumers in a
    community or ecosystem during a given period of time _____
    _____.

14.. Accumulation of net primary production is _____.

15. The ratio of belowgound to aboveground biomass is the _____ _____.

16. Organisms that feed on living organic matter are _____; organisms
    that live on dead organic matter are _____.

17. A turkey vulture feeding on a deer carcass is an example of a _____.

18. A rabbit feeding on soft fecal pellets is an example of _____.

77

19. Organisms in the food web that that reduce organic matter to inorganic molecules are _____.

20. A plant builds up stems and leaves that aid in the acquisition of more energy and nutrients through _____.

21. That part of photosynthate built up in the plant for future growth and other functions is _____.

22. In plants the increase in compounds, such as starch, that do not directly support growth, is called _____.

23. The synthesis of storage compounds from resources that would otherwise would be allocated to growth is _____ _____.

24. Plant move compounds that otherwise would be lost through litterfall from aging tissues to new growth by _____.

25. The point at which light received by phytoplankton is just enough to meet respiratory need and production equals respiration is the _____ _____.

26. Biomass gain by consumer organisms is _____ _____.

27. The movement of energy and nutrients from one feeding group to another that begins with plants and ends with carnivore, detrital feeders, and decomposers is a _____ _____.

28. Interacting food chains make up a _____ _____.

29. In consumers heat required for metabolism above basal or resting metabolism is _____ _____.

30. Energy remaining after plant respiration and stored as organic matter is _____ _____ _____.

31. Grazing mammalian herbivores, such as cattle, with a highly complex digestive system consisting of a four-compartment stomach are _____.

32. Organisms that depend upon on plant material as food are _____,

33. A robin feeding on an earthworm is functioning as a _____.

34. A white-footed mouse that eats seeds as well as beetles is an _____.

35. Insect larvae that slowly consume their host and then transform into another stage of their life cycles are _____.

36. A _____ _____ _____ begins at the producer level.

37. Plants that draw their nourishment from dead plant and animal material are
    _____.

38. An organism that spends part of its life cycle drawing its nourishment from another
    living organism is a _____.

39. _____ _____ is a functional classification of organisms according to
    feeding relationship from first level autotrophs through succeeding levels of
    herbivores and carnivores.

40. Three types of ecological pyramids are the _____ ___ _____, based on
    decreasing number of organisms with each successive trophic level; the _____
    _____, based on the decreased weight of organisms on each successive
    trophic level; and the _____ ___ _____ which is based on the
    quantity of energy fixed, stored and passed along each successive trophic level.

41. A method of measuring primary production that measures the consumption of and
    production of oxygen under light and dark conditions is the _____ and
    _____ _____ _____.

42. Species in a food web that are trophically similar to the same prey and same
    predators are called _____.

43. A technique of measuring production of terrestrial ecosystems involving removal
    and weighing vegetation at periodic intervals is the _____ _____,

44. A modified version used in forest ecosystems involves measurements of height,
    weight and other factors is called _____ _____.

45. A method of measuring primary productivity by calibrating the content of chlorophyll
    in samples of plants in an ecosystem is the _____ _____
    method.

46. A technique for measuring primary productivity by monitoring changing carbon
    dioxide concentrations as a means of determining the rate of carbon dioxide uptake
    is the _____ _____ _____ method.

Self Test

True and False

1. _____ A mouse has a higher assimilation efficiency than a salamander.

2. _____ If a deer consumes 1000 kcal of plant energy, it would convert about 10
   kcal into herbivore tissue.

3. _____ The major pathway of energy in terrestrial ecosystems is the detrital food
   chain.

4. _____ The rate at which plants convert solar energy into chemical energy of organic molecules is gross primary production.

5. _____ GPP - R = NPP

6. _____ The trophic level of an organism refers to the number of steps it is away from primary production.

7. _____ Organisms that feed on dead organic matter are biophages.

8. _____ Biomass accumulates when P = R.

9. _____ Mice have a greater growth efficiency than grasshoppers.

10. _____ Production efficiency in plants is the ratio of net primary production (NPP) to gross primary production (GPP).

11. _____ A carnivore that feeds on a herbivore would be a second-level consumer.

12. _____ Energy is the ability to do work.

13. _____ Photosynthesis is an exothermic reaction.

14. _____ The region of maximum productivity in a forest is at the top of the canopy.

15. _____ The digestive process in ruminants involves bacterial fermentation.

16. _____ The productivity of the open sea equals that of a temperate hardwood forest.

17. _____ The harvest method of estimating primary production estimates gross primary production.

18. _____ Standing crop is the amount of biomass per unit area per given time.

Matching

Productivity of Earth varies. Indicate the relative productivity of the following as **high**, **medium** or **low**.

1. _____ Deep ocean   5. _____ Estuaries, coastal waters

2. _____ Desert   6. _____ Temperate hardwood forest

3. _____ Moist grassland 7. _____ Modern agriculture

4. _____ Coniferous forest 8. _____ Tropical rain forest

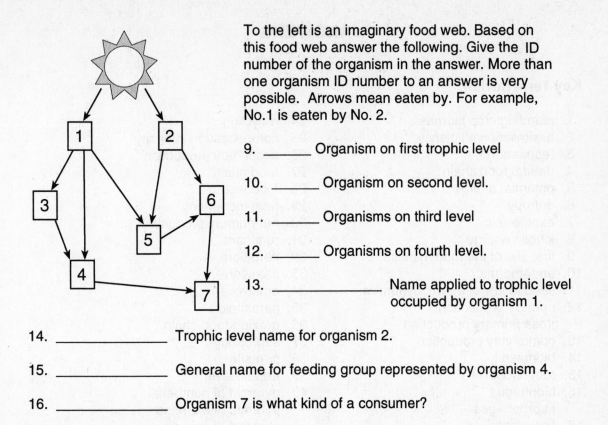

To the left is an imaginary food web. Based on this food web answer the following. Give the ID number of the organism in the answer. More than one organism ID number to an answer is very possible. Arrows mean eaten by. For example, No.1 is eaten by No. 2.

9. _____ Organism on first trophic level

10. _____ Organism on second level.

11. _____ Organisms on third level

12. _____ Organisms on fourth level.

13. _____ Name applied to trophic level occupied by organism 1.

14. _____ Trophic level name for organism 2.

15. _____ General name for feeding group represented by organism 4.

16. _____ Organism 7 is what kind of a consumer?

Short Answer Questions

1. What is the difference in assimilation and energy allocation between homeotherms and poikilotherms?

2. What is the ecological significance of root/ shoot ratios?

81

ANSWERS

Key Term Review

1. standing crop biomass
2. assimilation efficiency
3. respiration
4. detrital food chain
5. potential energy
6. entropy
7. exothermic
8. kinetic energy
9. first law of thermodynamics
10. endothermic
11. second law of thermodynamics
12. primary production
    gross primary production
13. community production
14. biomass
15. root/shoot ratio
16. biophages
    saprophages
17. scavenger
18. corprophagy
19. decomposers
20. growth
21. storage
22. accumulation
23. carbon dioxide flux
24. recycling
25. compensation intensity
26. secondary production
27. food chain
28. food web
29. heat increment
30. net primary production
31. ruminant
32. herbivore
33. carnivore
34. omnivore
35. parasitoid
36. grazing food chain
37. saprophage
38. parasite
39. trophic level
40. pyramid of numbers
    pyramid of biomass
    pyramid of energy
41. light and dark bottle method
42. trophic species
43. harvest method
44. dimension analysis
45. chlorophyll concentration
46. carbon dioxide flux

Self Test

True and False

1. T	10. T
2. F	11. F
3. T	12. T
4. F	13. F
5. T	14. F
6. T	15. T
7. F	16. F
8. T	17. F
9. F	18. T

Matching

1. low	9. 1, 3
2. low	10. 2, 4, 5, 6.
3. medium	11. 4, 6, 7
4. medium	12. 7
5. high	13. 1st level consumer
6. medium	14. herbivore
7. high	15. omnivore
8. high	16. carnivore

Short Answer Questions

1. Homeotherms, being endothermic, have a high metabolic rate necessary to maintain homeostasis. Most of the energy intake goes to meet maintenance costs and the remainder is allocated to growth. Homeotherms have a high assimilation efficiency. Poikilotherms, being ectothermic, allocate a minimal amount to homeostasis, which allows them to use most of their energy intake for growth. However, they have a low assimilation efficiency, which means much of their food intake passes off as waste.

2. Root/shoot ratios indicate a plant's allocation to belowground and aboveground production. In environments that are cold or low in nutrients or where competition for moisture and nutrients are high, plants allocate more of their net production to roots and other supporting structures. Because they have less biomass above ground, plants with high root/shoot ratios have lower net production.

CHAPTER 12
BIOGEOCHEMICAL CYCLES

Chapter Outline

Learning Objectives

After completing this chapter you should be able to:
- Describe biogeochemical cycling.
- Distinguish between gaseous and sedimentary cycles.
- Describe the carbon cycle and its relationship to potential global warming.
- Explain how the nitrogen cycle is driven by microbial processes.
- Describe the sulfur cycle, emphasizing its gaseous and sedimentary components.
- Describe the phosphorus cycle as an example of a purely sedimentary cycle.
- Discuss acid deposition, its types, sources, and effects on ecosystems.
- Show how chlorinated hydrocarbons introduced into ecosystems follow the biogeochemical cycle.
- Discuss cycling of radionuclides in aquatic and terrestrial ecosystems.

Summary

After reading the chapter and before continuing with the material below, read the Chapter Summary on pages 221-222. You will find related material on the following pages: solar radiation, 36-37, 101-102; atmospheric circulation, 39-42; ocean gyres, 42; water molecule, 64-65; water cycle, 67-70; nutrients, 117; micronutrients and macronutrients, 116; nutrient flows, 116-120; soil profile, 130-131; soil chemistry, 133-136; soil organic matter, 139-140; decomposition, 159-165; mineralization, 162; C:N ratio, 162-164; seasonal overturn in lakes, 290-293; nutrient spiraling in streams, 309-310; peatlands, 317; nitrogen cycling: grassland, 236; savanna, 245-246; desert, 249; tundra, 260-261; forest, 270-272, 278-280, 286; mutualism, 582-596.

Study Questions

1. What are the two types of biogeochemical cycles and what are their distinguishing characteristics? (196, 210)
2. What is the relationship between oxygen and life? (197)
3. What are the two major sources of oxygen in the atmosphere? (197)
4. Characterize the role of ozone in the stratosphere and in the atmosphere. Why is one the antithesis of the other? (198-199)
5. What is the role of chloroflurocarbons in the depletion of stratospheric ozone? (198-199)
6. What is the significance of the carbon cycle? (248-251)
7. What pools of CO_2 are involved in the global carbon cycle? (200-202)
8. Contrast the carbon pool of the ocean with that of the land relative to storage and turnover. (201-202)
9. What is the impact of burning fossil fuels on the carbon cycle? (203-205)
10. What is the greenhouse effect? (205-206)
11. What are some predicted potential outcomes of global warming? (206)
12. Characterize the following processes in the nitrogen cycle: fixation, ammonification, nitrification, and denitrification. (206-207)
13. What biological and nonbiological mechanisms are responsible for nitrogen fixation? (206-207)
14. What are sources of input into the nitrogen cycle? Sources of output? (206-208)
15. What are some major forms of nitrogen pollution? What are their sources? (208-210)
16. How does nitrogen pollution relate to ozone production in the atmosphere? (198, 209)
17. What is the source of sulfur in the sulfur cycle? Why does the sulfur cycle have characteristics of both sedimentary and gaseous cycles? (210-211)
18. In what manner does sulfur become a serious pollutant in the atmosphere and in aquatic systems? (210-212)
19. What are some effects of sulfur pollution? (210-212)
20. What is the source of phosphorus in the phosphorus cycle? The sink? (212-213)
21. Through what three compartments does the phosphorus cycle move in aquatic ecosystems? (213, Figure 12.20)
22. How have human activities altered the phosphorus cycle? (213)
23. What are the major sources of lead pollution? What are the effects of such pollution? (213-214)
24. What are the ecological effects of acid deposition on terrestrial and aquatic ecosystems? (214-216)
25. What is acid deposition? Distinguish between dry deposition and wet deposition. (214)
26. What are the major constituents of acid deposition? (214-215)
27. What is the relationship of acid deposition to forest decline in the Appalachians and in the mountains of Europe? What mechanisms might be involved? (216)
28. How are DDT, PCB, and other chlorinated hydrocarbons cycled in ecosystems? (216-217)
29. What effects do DDT and PCB have in terrestrial and aquatic ecosystems? What organisms are most seriously affected? (218-219)
30. What are the sources of radionuclides in ecosystems? (219)

31. What is the most common route of dispersion of nuclides in terrestrial ecosystems? (219; Figure 12.25, 12.26)
32. How does dispersion in aquatic ecosystems differ from terrestrial ecosystems? (220)
33. Do all radionuclides concentrate in higher levels in food chain? (220)

Key Terms and Phrases

biogeochemical cycle
greenhouse effect
fixation
nitrification
ammonification
denitrification
gaseous cycle

sedimentary cycle
eutrophication
dry acid deposition
wet acid deposition
ionizing radiation
radionuclides

Key Term Review

1. High energy, short wavelength radiation capable of removing some electrons from atoms and attracting them to others is known as _____ _____.

2. Fission and nonfission products released to the environment by nuclear testing and nuclear wastes are _____.

3. The conversion of gaseous nitrogen to ammonia and nitrates by biological and abiotic means is _____.

4. The nutrient enrichment of freshwater ecosystems is _____.

5. Acid pollutants that return to Earth as particulate matter and air-borne gases is _____ _____ _____.

6. Bacterial oxidation of ammonia to nitrates and nitrites is _____.

7. Acid pollutants carried to Earth by rain, fog, and snow is _____ _____ _____.

8. The _____ _____ consists of two phases, salt solution and rock.

9. Mineralization of organic nitrogen is _____.

10. Nutrients whose main pools are in the atmosphere follow a _____ _____.

11. Chemical exchanges of elements among atmosphere, Earth's rocks, and living things make up _____ _____.

12. Atmospheric gases, especially carbon dioxide and methane, that reradiate longwave radiation back to Earth produce a _____ _____.

13. The process of bacterial conversion of nitrates to atmospheric nitrogen is _____

Self Test

True and False

1. _____ The major global sink for CO_2 is the ocean.

2. _____ Calcium is an example of a sedimentary biogeochemical cycle.

3. _____ The largest reservoirs of carbon in the global ecosystem are forests.

4. _____ CO_2 concentration in the atmosphere is increasing at an exponential rate.

5. _____ The rate of CO_2 production by respiration is nearly three times higher in summer than in winter.

6. _____ Cyanobacteria are important nonsymbiotic nitrogen fixers.

7. _____ A major sink of phosphorus in aquatic ecosystems is open water.

8. _____ Nitrogen is involved in the eutrophication of freshwater ecosystems.

9. _____ A major impact of acid deposition in aquatic systems is the release of aluminum ions.

10. _____ Chlorinated hydrocarbons, such as DDT, accumulate in the fat of animals.

11. _____ Plants absorb radionuclides contaminants through their foliage.

12. _____ Concentrations of radionuclides increase consistently through an aquatic food chain.

13. _____ The least active reservoir for carbon is terrestrial.

14. _____ If level of phosphorus were increased in a lake, it would experience an algal bloom.

15. _____ Highly concentrated amounts of pesticides are used around the American home.

16. _____ The interactions of automobile and industrial pollutants with ultraviolet light result in several secondary pollutants.

17. _____ Acid deposition kills plants on contact.

18. _____ Injection of chloroflurocarbons and nitrogen oxides into the atmosphere causes destruction of stratospheric ozone.

Matching

Match these terms with the statements given below: Terms may be used more than once.

A. SO_2 D. Phosphorus   G. CO_2 J. Calcium
B. NO_2 E. Lead   F. DDT   H. PAN   K. Methane
C. Cesium-137   F. DDT   I. Sulfur

1. _____ Zooplankton excrete and recycle this element in aquatic ecosystems.

2. _____ Excessive quantities of this material released into the atmosphere are implicated in acid deposition.

3. _____ Excessive quantities of this material released into the atmosphere appear to cause overfertilization and death of coniferous forests at high elevations.

4. _____ DDT and other chlorinated hydrocarbons mimic this element in their passage through the ecosystem and food chain.

5. _____ The biogeochemical cycle of this element is both sedimentary and gaseous.

6. _____ This pollutant breaks down to eventually form ozone.

7. _____ Fallout of this pollutant came from auto exhausts.

8. _____ Causes thinning of egg shells of birds.

9. _____ Injected into the atmosphere by the burning of fossil fuel, some of it is scrubbed from the atmosphere by plant growth.

10. _____ This radionuclide accumulates in muscle tissue of animals rather than bone.

11. _____ This material is one of the greenhouse gases.

Short Answer Questions

1. The sedimentary cycle consists of a rock phase and a salt solution phase. Why is the salt solution phase so important in the sedimentary cycle?

2. Briefly outline the phosphorus cycle in aquatic ecosystems.

3. Explain how nitrogenous atmospheric pollution in part may relate to the decline of coniferous forests at high elevations in eastern North America and Europe.

ANSWERS

Key Term Review

1. ionizing radiation
2. radionuclides
3. fixation
4. eutrophication
5. dry acid deposition
6. nitrification
7. wet acid deposition
8. sedimentary cycle
9. ammonification
10. gaseous cycle
11. biogeochemical cycles
12. greenhouse effect
13. denitrification

Self Test

True and False

1. T	10. T
2. T	11. T
3. F	12. f
4. T	13. F
5. T	14. T
6. T	15. T
7. F	16. T
8. T	17. F
9. T	18. T

Matching

1. D	7. E
2. A/B	8. F
3. B	9. G
4. J	10. C
5. I	11. K
6. H	

Short Answer Questions

1. The sedimentary cycle consists of two phases, rock and salt solution. In the rock phase the nutrients are locked in rock and sediments. They are released by weathering. Only when they become involved in part of the water cycle and go into solution are these nutrients available to autotrophs, microflora, and bacteria. From this point, the nutrients can move through food webs.

2. Phosphorus in aquatic ecosystems consists of four forms: particulate phosphorus, inorganic phosphorus, organic phosphorus secreted by zooplankton, and colloidal phosphorus derived form organic phosphorus. The two important forms are particulate phosphorus and inorganic phosphorus. The organic and colloidal phosphorus release phosphate to the inorganic fraction that is rapidly recycled through phytoplankton. Phosphorus in phytoplankton is ingested by zooplankton. Zooplankton excrete phosphorus, over one half of which is inorganic phosphorus that is taken up quickly by phytoplankton.

3. Atmospheric nitrogenous pollution results in the deposition of nitrogen. This deposition occurs in nitrogen-limited ecosystems such as the high altitude and northern forests. Because of their nitrogen-limited environment, these forests are efficient at retaining and recycling nitrogen from precipitation and organic matter. Because of nitrogen deposition, these forests receive more nitrogen than they can handle. Increased growth stimulated by N fertilization continues into summer, only

to be killed back during the winter. Tree growth exceeds the availability of other needed nutrients and begin to show symptoms of other nutrient deficiencies. As the nitrogen level increases, root biomass decreases, because sufficient nitrogen is close by. Trees are not stimulated to increase root growth to reach new supplies. The resulting smaller root biomass inhibits the uptake of other nutrients and water. These conditions appear to lead to decline and death of trees. It is a case of too much of a good thing is harmful.

PART 4

COMPARATIVE ECOSYSTEM ECOLOGY

CHAPTER 13
GRASSLAND TO TUNDRA

Learning Objectives

After completing this chapter you should be able to:
- Describe the types of grassland.
- Summarize the structure of grassland ecosystems.
- Discuss the relationship of grasslands to precipitation.
- Describe the role of grazing in grassland ecosystems.
- Contrast belowground and aboveground primary production in grasslands.
- Discuss the major features of the nitrogen cycle in grasslands.
- Describe the characteristics of savanna vegetation.
- Discuss the nitrogen cycle in savannas.
- Describe the types of shrublands.
- Explain the unique features of shrublands.
- Discuss the function of shrublands with emphasis on mediterranean-type ecosystems.
- Describe the characteristics of desert ecosystems.
- Explain how desert plants are adapted to arid environments.
- Discuss the relationship of precipitation to primary production in the desert.
- Contrast arctic and alpine tundra ecosystems.
- Discuss the relationship between frost and the physical structure of the tundra.
- Explain how primary production and nutrient cycle are influenced by permafrost and the short growing seasons.

Summary

After reading the chapter and before continuing with the material below, read the Chapter Summary on pages 261-262. You will find related material on the following pages: climate 36-48; microclimate, 48-53; climate and vegetation, 54-60; water use efficiency, 70-71; plant adaptation to aridity; 71-72 animal adaptations to aridity, 75-76; adaptations to cold, 82. 84-85, 88-89; 92-95; nutrient cycling, 116-120; soil development, 140-143; C_4 plants, 154-155; decomposers, 159-165; herbivores, 181, 183-184; nitrogen cycle, 206-210; population cycle, 403-406; 543-545; plant defenses, 524-529; physical structure of communities, 598-601; grazing disturbance, 645-646; succession, 656-677.

Study Questions

1. What characteristics do all grasslands have in common? (226)
2. Distinguish between a bunch grass and a sod grass. (226)
3. Traveling east to west across North America what are (or were) the five major natural grassland formations and what are the distinguishing characteristics of each? (226-230)
4. What are the distinguishing characteristics of cultivated, annual, and successional grasslands? (227, 230-231)
5. What are the seasonal changes in grassland structure? (231)
6. What forms of animal life are characteristic of grasslands and what are some of their behavioral adaptations? (232)
7. Discuss net production in grasslands. Where is net production the greatest within a grassland? (233-234, Table 13.1)
8. What is the impact of grazing on grasslands? How does above-ground herbivory compare to below-ground herbivory. (233-236)
9. What distinguishes savannas from grasslands both in structure and function? Under what climatic conditions have they evolved? (237)
10. Describe the role of fire and moisture in savanna ecosystems? (237)
11. What is the role of trees, shrubs, and grass to the horizontal structure of savannas? (237)
12. What is the relationship between trees, grass, and nitrogen cycling in savanna ecosystems? (238)
13. What features are unique to shrublands? (240)
14. What are mediterranean-type shrublands? What is their relationship to fire? (240-242)
15. What are heathlands and where do they occur? (242-244)
16. What is the relationship between rainfall patterns and nitrogen cycle in mediterranean-type ecosystems? (245-246)
17. What climatic forces lead to deserts? (246-247)
18. What are the major features of deserts? (247)
19. What is a cool desert? A hot desert? (247)
20. How are plants adapted to survive in the desert? What are drought resisters? Drought evaders? (247)
21. Upon what does primary production in the desert depend? (248)
22. Why is nitrogen limiting in desert ecosystems? (249)
23. What is the relationship between grazing and desert vegetation? (249)

24. Why are herbivores and carnivores opportunist rather than specialist feeders in the desert ecosystem? (249-250)
25. What is the importance of granivores to desert ecosystems? (249)
26. What physical and biological features characterize the tundra? (250-251)
27. Define frost hummock, stone polygon, solification terrace, and permafrost. (251-252)
28. Contrast adaptations of plants to light in arctic and alpine tundras. Why the difference? (252)
29. What is the Krummholz? (253)
30. Contrast the structural features of the arctic tundra with those of the alpine tundra. (255-257)
31. What is the relationship between tundra vegetation and permafrost? (251, 255)
32. What are two dominant herbivores and two dominant carnivores on the arctic tundra? (255)
33. Where is the bulk of primary production in the arctic tundra? Why? (258)
34. What is the role of grazing herbivores in the arctic tundra? (259-260, 261)
35. What is unique about the nitrogen cycle in the arctic tundra? (260-261)

Key Terms and Phrases

grassland	garrigue	chaparral	granivores
savanna	maqui	heathland	tundra
desert	mattoral	bajadas	fellfield
shrubland	fynbos	playas	Krummholz
successional shrubland	mallee	arrayos	permafrost
cyroplanation			

Key Term Review

1. Permanent frozen area of ground in the Arctic is known as
   _____.

2. A _____ is a tropical grassland with scattered trees and shrubs.

3. The _____ is the zone of stunted trees at timber line.

4. _____ are ecosystems dominated by multiple-stemmed, relatively low woody vegetation.

5. In the tundra bare, rock-covered ground on high exposed areas is called a
   _____.

6. Molding of tundra landscape by frost is _____.

7. Australian shrubland dominated by Eucalyptus is known as _____.

8. In North America, mediterranean-type vegetation dominated by schlerophyllic evergreen vegetation is called _____. In the Mediterranean region of Europe its ecological equivalent is known as _____.

9. In the Mediterranean region of Europe, two types of shrublands are the result of disturbance. The degradation of pine forests produced the _____, and the _____ replaced cork oak forests.

10. Shrub communities dominated by ericaceous shrubs make up _____. In South Africa such shrublands are called the _____.

11. Erosion of desert landscape produces large areas of debris known as _____. Water-cut desert canyons are _____, and low basins that receive water from the canyons and hills are called _____.

12. _____ _____ are temporary types of ecosystems that eventually give way to forests.

13. _____ are areas in arctic and alpine regions characterized by bare ground, absence of trees, and growth of mosses, lichens, sedges, forbs, and low shrubs.

14. Animals that feed largely on seeds are _____.

Self Test

True and False

1. _____ Grasslands of arid regions use water efficiently.

2. _____ Belowground biomass in grasslands is greater than aboveground biomass.

3. _____ Grasses have a high root/shoot ratio.

4. _____ Some grasses are more sensitive to grazing than others.

5. _____ Moderate grazing stimulates primary production in grasslands.

6. _____ Major decomposers in grassland ecosystems are bacteria.

7. _____ Accumulation of mulch is central to nutrient cycling in grassland.

8. _____ A large standing crop of mulch is detrimental to nitrogen cycles in grassland.

9. _____ Fire plays a major role in the maintenance of grasslands.

10. _____ Grassland strata change with the seasons.

11. _____ The major structural element in savannas is sod grasses.

12. _____ Dominant herbivores in South American savanna are invertebrates.

13. _____ Most nitrogen in savanna trees goes into root reserves.

14. _____ Savannas have a well developed horizontal structure.

15. _____ Nitrogen cycling in savannas is highest beneath trees.

16. _____ Shrubs comprise a distinct taxonomic category.

17. _____ Shrubs in mediterranean-type ecosystems are fire-adapted.

18. _____ Successional shrub ecosystems are important wildlife habitat.

19. _____ Desert topography is shaped by water.

20 _____ Dominant plants in North America cool deserts are cacti.

21. _____ Warm desert plants have a high root/shoot ratio.

22. _____ Drought evasive plants survive the dry period as seeds.

23. _____ Ephemeral desert communities annually experience 100% turnover of aboveground and belowground biomass.

24. _____ Desert plants tend to retain nutrients in biomass before shedding leaves.

25. _____ Granivores can consume most of the desert plants' seed production.

26. _____ If rains fail, some desert animals do not reproduce.

27. _____ Omnivory rather than carnivory rules in desert ecosystems.

28. _____ The tundra is characterized by unique types of vegetation.

29. _____ Most net production of arctic tundra enters the detrital food chain.

30. _____ Of all ecosystems, the arctic tundra has the smallest proportion of nutrient capital in aboveground biomass.

31. _____ Leaching of nutrients is high in tundra ecosystems.

32. _____ Alpine tundras are found in tropical regions.

33. _____ Alpine tundra plants require longer periods of daylight than arctic tundra plants.

34. _____ Frost and alternate freezing and thawing mold the tundra landscape.

35. _____ Alpine plants reach higher light saturation points than arctic tundra plants.

36. _____ Permafrost impedes the drainage of water and creates wet conditions in dry parts of the arctic tundra.

37. _____ Continuous turnover is essential in the replenishment of nitrogen and phosphorus in arctic tundra ecosystems.

38. _____ Lemmings and other tundra have little influence on nutrient cycling in the arctic tundra.

Matching

Associate each of the following ecosystems with their appropriate climatic conditions.

A. Desert B. Mediterranean-type shrubland C. Tundra D. Grassland E. Savanna

1. _____ Rainfall frequent; periodic droughts; evaporation rapid.

2. _____ Extreme seasonal fluctuations in rainfall; warm to hot temperatures.

3. _____ Hot dry summers; cool moist winters.

4. _____ Low precipitation, low temperatures.

5. _____ Evaporation exceeds rainfall.

Associate each of the following ecosystems with their appropriate process of cycling nitrogen.

A. Savanna B. Grassland C. Tundra D. Australian mallee E. Desert

6. _____ Plants lower N concentration of surrounding soil; withdraw much N from leaf into stem before litterfall and concentrate the rest of N though litterfall directly beneath plants.

7. _____ Trees move N into root growth prior to leaf fall, where it is available for flush of new seasonal growth .

8. _____ Ninety percent of N is tied up in soil organic matter. N translocated to roots is moved up to new growth. Turnover rapid. N in green plants one year reenters new plants the following year.

9. _____ N cycling light; most N tied up in living and dead plant biomass. N collects in litter beneath plant; N-fixing cyanobacteria concentrated beneath shrubs adds to N supply.

10. _____ Most N accumulates and is stored in living and dead plant material where it is unavailable for recycling. Constant turnover of N between exchangeable and soluble pools is needed.

Short Answer Questions

1. Describe two ways in which desert plants evade drought.

2. Contrast arctic and alpine tundras relative to environmental conditions and physiological responses.

3. How does competition influence the distribution of trees on the African savanna?

4. Why do shrubs have a competitive advantage over grass?

5. What is the role of mulch in a grassland ecosystem?

ANSWERS

Key Term Review

1. permafrost
2. savanna
3. Krummholz
4. shrublands
5. fellfield
6. cryoplanation
7. mallee

8. chaparral, mattoral
9. garrigue, maque
10. heathland, fynbos
11. bajadas, arroyos, playas
12. successional shrubland
13. tundras
14. granivores

Self Test

True and False

1.	F.	13.	T	25.	T	37.	F
2.	T	14.	T	26.	T	38.	F
3.	T	15.	T	27.	T		
4.	T	16.	F	28.	F		
5.	T	17.	T	29.	F		
6.	F	18.	T	30.	T		
7.	T	19.	T	31.	F		
8.	T	20.	F	32.	T		
9.	T	21.	F	33.	F		
10.	T	22.	T	34.	T		
11.	F	23.	T	35.	T		
12.	T	24.	T	36.	T		

Matching

1. D
2. E
3. B
4. C
5. A
6. D
7. A
8. B
9. E
10. C

Short Answer Questions

1. Some desert plants are drought-evaders. Some persist as seeds during the dry periods and germinate, sprout, and flower when sufficient moisture becomes available. Some woody plants have deep roots that reach the water table. Some drought resisters shed leaves during dry periods; some store water in their tissues; others have very shallow root systems capable of picking moisture quickly from light rains.

2. The arctic tundra has a short growing season and very long days; the alpine tundra has a longer growing season, shorter days, high light intensity, and strong winds. Thus alpine plants have a high light saturation point, and very low cushion growth forms that can withstand winds. Sedges and cotton grass are restricted to protected areas. Cushion plants are much less conspicuous in arctic tundra and cotton grasses and sedges are more common plant communities. Arctic tundra plants reproduce largely by vegetative means where as alpine tundra plants reproduce by seeds.

3. Grasses and trees compete for available moisture, but most intense competition takes place among trees. Competition for moisture by tree roots results in wide spacing between trees.

4. Shrubs have extensive root systems that can draw moisture deep in the soil. Their high root/shoot ratio allows shrubs to put more nutrients into root biomass rather than aboveground biomass, increasing their efficiency in nutrient uptake. Their perennial woody growth immobilizes nutrients, slowing nutrient cycling and allowing invasion of shrubs into grassland.

5. Mulch conserves soil moisture by reducing evaporation, but too much mulch retards infiltration of water into the soil. By insulating soil from solar radiation, mulch inhibits activity of soil microbes and reduces root productivity. Grazing and fire reduce excessive accumulation of mulch.

CHAPTER 14
FORESTS

Chapter Outline

Coniferous Forests
   Types
   Structure
   Function
Temperate Broadleaf Forests
   Types
   Structure
   Function

Tropical Forests
   Types
   Structure
   Function
   Fate of Tropical Forests
Summary

Learning Objectives

After completing this chapter you should be able to :
- Describe types of coniferous, temperate, and tropical forests.
- Discuss the three major growth forms of coniferous trees.
- Contrast nutrient cycling in old-growth and young-growth coniferous forests.
- Describe the nitrogen cycle in the canopy of old growth forest.
- Discuss the role of dead wood in the forest.
- Explain the influence of vertical structure of the forest on microclimate.
- Outline the major features of nutrient cycling in temperate deciduous forests.
- Explain the role of mycorrhizae in the nutrient cycling in coniferous and tropical forests.
- Contrast nutrient cycling among coniferous, deciduous, and tropical forests.
- Discuss the impact of deforestation in the tropics.

Summary

After reading the chapter and before continuing with the material below, read the Chapter Summary on page 288. You will find related material on the following pages: relative humidity, 45; Holdridge life zone, 6, 58-59; light, 100; internal nutrient cycling, 116-120; soil chemistry, 133-136; mor humus, 139; tropical soils, 140-41; soil groups, 144-146; decomposition, 159-165; primary production, 169-175; nitrogen cycle, 210-212; phosphorus cycle, 212-213; mycorrhizae, 528, 584-585; mutualism, 582-587.

Study Questions

1. Discuss the differences and similarities of the several broad types of coniferous forests and their geographical locations. How are they influenced by climate? (264-266)
2. Contrast nutrient cycling in a young and old growth Douglas-fir forest. (269-70)
3. Contrast nutrient cycling in southern loblolly pines with Douglas-fir forests. (271)
4. What are accumulator plants? (271)
5. Describe nitrogen cycling by nutrient-scavenging microcommunities in the canopy of Douglas fir. (271)
6. What is the role of litter-decomposing fungi in coniferous forest ecosystems? How do they act as nutrient sinks? (272)
7. What is the role of dead wood in a forest ecosystem? (272-273)
8. What are the major types of temperate broadleaf forests? (272-273)
9. What are the three major forest types within the central hardwood forests? (274)
10. Describe the vertical structure of a hardwood forest and its influence on forest microclimate.
11. Outline the major features of nutrient cycling in a deciduous forest. (278-279)
12. Contrast the biological cycling of nutrients in a deciduous forest and a coniferous forest. (280)
13. Distinguish among tropical rain forest, tropical seasonal forest, and tropical dry forests. (281-282)
14. Describe the structural features of a tropical rain forest. (283-285)
15. How do these structural features reality to stratification of animal life? (284-285)
16. What is the close interaction between animals and plant life in the tropical rain forest? (284-285)
17. What is the change in growth patterns of tropical rainforest trees as they grow in height? What is crown-shyness? (283)
18. Outline nutrient cycling in tropical rain forests. What is the role of mycorrhizae? (285-286)
19. What is the major loss in the destruction of tropical rain forests? (287)

Key Terms and Phrases

old growth forest	epiphytes	tropical dry forest
taiga	emergents	lianas
cove forest	crown shyness	jungle
rain forest	tropical dry forest	

Key Term Review

1. Evergreen forests rich in epiphytes and lianas of the permanently wet tropics are _____ _____.

2. Forests that have not been disturbed by humans for hundreds of years and are characterized by accumulation of dead wood are _____ _____ _____.

3. _____ is a name given to open boreal forests encircling the Northern Hemisphere.

4. Rich mesophytic forests growing on the northern and eastern slopes of the southern Appalachian mountains of eastern North America are called _____ _____.

5. Plants, such as orchids, that use another plant for support but do not draw nourishment from it are _____.

6. Woody, free swinging, climbing plants are _____.

7. Dense secondary tropical forest coming in where tropical rain forest has been disturbed creates a _____.

8. Individual trees that stand above the canopy of lowland tropical forests are called _____.

9. Tropical regions with a marked dry season support tropical _____ forests; tropical regions with a very long dry season, during which trees and shrubs lose their leaves, support tropical _____ forests.

10. Crowns of tropical rain forest trees that fit together like widely spaced jigsaw puzzles are growth patterns termed _____ _____.

Self Test

True and False

1. _____ Southern pine forests are a successional stage leading to a temperate deciduous forest.

2. _____ Vertical structure is well developed in coniferous forests.

3. _____ Pollen rain in a spruce forest is a nutrient source for fungi.

4. _____ Epiphytes obtain nutrients from tree trunks.

5. _____ Dead wood is a critical component of forest ecosystems because it provides foraging areas and nesting sites for many birds and mammals.

6. _____ Fungi are nutrient sinks in coniferous forests.

7. _____ Sun flecks influence the distribution of herbaceous plants on the forest floor.

8. _____ Mineral cycling in the forest can be maintained only if nutrients are withdrawn from the soil.

9. _____ Litter is the most important nutrient pool in the temperate deciduous forest because the nutrients can be recycled rapidly.

10. _____ Much of the nutrient pool in a forest is stored in biomass unavailable for short term recycling.

11. _____ Coniferous forests are more efficient at recycling nutrients than deciduous forests.

12. _____ Animals in the tropical rain forest are essential for pollination and seed dispersal.

13. _____ Conifers can remove elements from the soil in quantities great enough to upset nutrient balance in the ecosystem.

14. _____ Young tropical forest trees are monopodial, that is they consist of a single stem and a tall narrow crown.

15. _____ The litter layer in a coniferous forest is thick but its decomposition is fast.

16. _____ Tropical forest plants depend heavily on animals for pollination of flowers and dispersal of seeds.

Matching

Match the coniferous forest type with its characteristic forest tree.

1. _____ Taiga   A. Loblolly pine

2. _____ Temperate rain forest   B. Redwoods

3. _____ Montane coniferous forest   C. Black spruce

4. _____ Woodland   D. Douglas-fir

5. _____ Southern pine forest   E. Pinon pine

If the item in Column A is greater than the item in column B, place **A** on the line. If the item in column B is greater than the item in column place **B** on the line. If they are equal, place a **C** on the line.

	A	B
1. _____	Living biomass in roots of young Douglas-fir forest.	Living biomass in roots of old-growth Douglas-fir forest.
2. _____	Detrital organic matter in litter of young Douglas-fir forest.	Detrital organic matter in litter of old-growth Douglas-fir forest.
3. _____	Detrital organic matter in soil of young Douglas-fir forest.	Detrital organic matter in soil of old-growth Douglas-fir forest.
4. _____	Retention of N in a coniferous forest.	Retention of N in a deciduous forest.
5. _____	Midsummer temperature in canopy of deciduous forest.	Midsummer temperature on floor of deciduous forest.
6. _____	Winter temperature in canopy of a deciduous forest.	Winter temperature in the litter layer of a deciduous forest.
7. _____	Heterotrophic respiration in cove forest stand.	Heterotrophic respiration in oak-pine forest stand.
8. _____	Nitrogen accumulation in roots of young loblolly pine.	Nitrogen accumulation in roots of young Douglas-fir.
9. _____	Woody biomass in temperate deciduous forest.	Woody biomass in tropical rain forest.
10. _____	Summertime humidity in canopy of temperate deciduous forest.	Summertime humidity at the floor of a temperate deciduous forest.
11. _____	Nitrogen sequestered in foliage of Douglas-fir.	Nitrogen sequestered in foliage of loblolly pine.
12. _____	Nutrient cycling through a coniferous forest.	Nutrient cycling through a deciduous forest.

Short Answer Questions

1. If tropical forest soils are so nutrient poor, how can tropical rain forests support very high biomass?

2. Describe the nitrogen cycle in the microcommunity in the foliage of old-growth Douglas-fir forests.

3. Why are coniferous forests unable to replace foliage loss by attacks of defoliating insects?

ANSWERS

Key Term Review

1. rain forest
2. old-growth forests
3. taiga
4. cove forests
5. epiphytes

6. lianas
7. jungle
8. emergents
9. seasonal, dry
10. crown shyness

Self Test

True and False

1. T	9. T
2. F	10. T
3. T	11. F
4. F	12. T
5. T	13. T
6. T	14. T
7. T	15. F
8. T	16. T

Matching

1. C
2. B
3. D
4. E
5. A

Greater Than, Lesser Than Questions

1. A	4. A	7. A	10. B
2. B	5. A	8. A	11. B
3. A	6. B	9. C	12. B

Short Answer Questions

1. In tropical forests, litter decomposes rapidly. Feeder roots of tropical trees are concentrated in the humus layer rather than in the upper mineral soil. These feeder roots with the aid of abundance of mycorrhizae take up nutrients rapidly, preventing the leakage of nutrients to mineral soil.

2. The canopy microcommunity supports producers, consumers, and decomposers. Lichens with cyanobacteria fix atmospheric nitrogen. Nitrogen lost from leaching combines with canopy moisture to form organic solutions taken up by microorganisms and epiphytes. Part of the microbial production is eaten by canopy arthropods.

3. Each fall deciduous forests translocate to and store nutrients and carbohydrates in the roots, from which they can be mobilized. Coniferous forests store most of their carbohydrates and nutrients in evergreen canopies. For this reason coniferous trees cannot mobilize resources to replace lost foliage.

Chapter 15
Freshwater Ecosystems

Learning Objectives

After completing this chapter you should be able to:

- Describe the seasonal physical stratification of a lake ecosystem and its influence on life in a lake.
- Discuss the structure of lentic ecosystems.
- Describe nutrient cycling and energy flow in lakes.
- Explain eutropy, oligotrophy, and dystrophy.
- Describe the physical characteristics of flowing water ecosystems.
- Explain the order classification of streams.
- Describe the problems of nutrient cycling in flowing water ecosystems.
- Discuss the sources of energy in lotic ecosystems.
- Discuss the functional role of stream organisms in the processing of organic matter in streams.
- Explain the role of spiraling in nutrient cycling in flowing water.
- Describe the river continuum concept and regulated rivers.
- Discuss the criteria needed to delineate a wetland.
- Describe types of wetlands.
- Explain hydroperiod and its role in structuring wetlands.
- Contrast nutrient cycling in freshwater marshes and a bog.

Summary

After reading the chapter and before moving on to the material below, read the Chapter Summary on pages 323-324. You will find related material on the following pages: structure of water, 64; specific heat of water, 65; water cycle, 65-69; plant responses to flooding, 73-75; light, 101-105; nutrients and nutrient cycles, 116-120; gley soil, 142-143; decomposition, 159-165; energy flow, 189-192; food webs, 192-193, 336, 621-623; oxygen cycle, 196-199; nitrogen cycle, 210-212; phosphorus cycle, 212-213.

Study Questions

1. What causes temperature stratification in lakes and ponds? (290)
2. What are the epilmnion, metalimnion and hypolimnion? What are their significance? (290)
3. What is the ecological significance of oxygen stratification in summer and winter? (291)
4. What is the spring and fall overturn and what is its ecological importance to life in a lake? (290)
5. What is compensation depth? (294)
6. What is the limnetic region of a lake and how is life adapted to it? (290-295)
7. Contrast the limnetic region with the littoral region of a lake. (294-295)
8. What contribution does the littoral zone make to a lake ecosystem? (297)
9. What role does detritus play in a lake ecosystem? (297)
10. Describe nutrient cycling and energy flow within a lake ecosystem. (297-298)
11. Contrast between oligotrophy, eutrophy, and dystrophy. (298-301)
12. What are the outstanding characteristics of running water ecosystems? (301-303)
13. What is the relationship of current to the nature of a stream. (303-305)
14. Contrast the function of riffles and pools in the stream ecosystem. (303)
15. In what way are stream organisms adapted to living in flowing water? How do adaptations change as fast streams become slow? (303-305)
16. What is the major source of nutrients in streams? (305)
17. What is the basic energy source of headwater streams? (303)
18. What is meant by DOM, FPOM, and CPOM? (305)
19. What is spiraling and how does it function in nutrient cycling in streams? What compartments and consumer groups are involved? (306-309)
20. What are the major invertebrate feeding groups in a stream? (307-308)
21. What is the river continuum? (309-311)
22. How do the downstream lotic systems relate to upstream systems? (311)
23. What are regulated rivers and how do they affect the river continuum? (311-312)
24. What indicators define a wetland? (313-316)
25. Distinguish between obligate and facultative wetland plants. (314-315)
26. Distinguish between a marsh, a swamp, and a bog. (316)
27. What is hydroperiod and what is its relationship to wetlands? (318)
28. What are the distinguishing characteristics of peatlands and what accounts for their uniqueness? (318-321)
29. Distinguish among blanket mire or moors, raised bogs, and quaking bogs. (317-318)
30. Distinguish between an ombrotrophic and a minerotrophic peatland. (322)
31. Contrast nutrient cycling in a marsh and a bog. (321-323)

Key Terms and Phrases

lotic	benthos	wetland
lentic	eutrophication	basin wetland
epilimnion	oligotropy	fringe wetland
metalimnion	biological oxygen demand	riverine wetlandl
thermocline	hypertrophic	marsh
hypolimnion	cultural eutrophication	swamp
overturn	dystrophic	riparian woodland

buffer system
trophogenic zone
tropholytic zone
littoral zone
emergent
limnetic zone
plankton
nekton
phytoplankton
zooplankton
benthic zone
uptake

marl lake
aufwuchs
basin
watershed
scrapers
collectors
grazers
shredders
spiraling
regulated river
profundal zone
turnover

vernal pool
peatland
mire
fen
bog
blanket mire
moor
raised bog
hydroperiod
quaking bog
hydrophytic

Key Term Review

1. Open water of a lake makes up the _____ _____.

2. _____ is the source of primary production in streams.

3. _____ describes nutrient poor aquatic ecosystems.

4. The warm, circulating upper layer of water in a lake in summer is the _____. Below it is the _____ which is characterized by the _____. The cold bottom layer is the _____.

5. _____ takes place when surface water cools to a point where temperature is uniform throughout the lake.

6. The _____ _____ comprises the bottom of a lake.

7. Nutrient enrichment of lakes is _____.

8. The _____ _____ contributes heavily to the input of organic matter in a lake.

9. _____ _____ develop along periodically flooded river banks.

10. Wetlands support _____ vegetation that includes pondweeds and cattails.

11. Flowing water ecosystems are termed _____ and ecosystems contained in a basin are termed _____.

12. The _____ _____ corresponds to epilimnion, the area of primary production, whereas the _____ _____ corresponds to the hypolimnion, the region of decomposition.

13. Organisms inhabiting the lake and pond bottom collectively are known as the _____.

14. Tiny plants that carry on photosynthesis in open water make up _____.

15. Feeding on these tiny plants are small animals known as _____.

16. Plants whose roots and lower stems are immersed in water and whose upper stems and leaves are above water are called _____.

17. Seasonal mixing of lake water in fall and spring is called _____.

18. Carbonate and bicarbonate ions form _____ _____ in lakes and streams that resist changes in pH.

19. Larger free-swimming organisms in aquatic ecosystems make up the _____.

20. The amount of oxygen needed for oxidative decomposition in aquatic ecosystems is the _____ _____ _____.

21. Accelerated nutrient enrichment of lakes by an influx of human and agricultural wastes is _____ _____.

22. Such high nutrient enrichments turns eutrophic lakes into _____ ones.

23. Lakes that receive large amounts of humic materials that stain the water brown are called _____.

24. Lakes that contain extremely hard water because of an input of calcium over a long period of time are called _____ _____.

25. Land area contributing to the flow for any one stream is its _____. A number of related basins make up a _____.

26. Functional groups in flowing water ecosystems include _____ that feed on leaves and larger bits of detritus; _____ that feed on algae growing on rocks. in streams; and _____ that filter out fine particulate organic matter.

27. The combined processes of nutrient cycling and downstream transport is called _____. In this process the distance an ion travels before being taken up is called _____. The distance it travels in particulate form before being released to dissolved organic matter is its _____ length.

28. Rivers whose flows are controlled by dams, levee, and channalization are _____.

29. Areas transitional between aquatic and terrestrial systems where water is at, near, or covering the land surface is one definition of a _____.

30. A _____ is a grassy wetland, a wooded wetland is a _____, and a _____ is dominated by Sphagnum moss.

31. Wetlands occupying depressions are _____ _____; those occupying coastal areas of larger lakes are _____ _____.

32. Extensive wooded tracts along rivers that are periodically flooded are _____ _____.

33. Shallow wetlands flooded in winter and spring, but dry in summer and fall are _____ _____.

34. Wetlands with considerable amounts of water retained by an accumulation of partially decayed organic matter are peatlands or _____.

35. Bogs that develop on uplands where a peat forms a barrier against downward flow of water are called _____ _____ or _____.

36. Peatlands fed by water flowing through mineral soils and dominated by sedges are _____.

37. _____ _____ result when a lake basin fills and the domed surface is raised above the influence of ground water. When a lake basin fills from above rather than from below, resulting in a floating mat of peat, a _____ _____ develops.

38. The structure of a wetland is influenced by its _____, which includes the duration, frequency, depth, and season of flooding.

Self Test

True-False

1. _____ The water at the bottom of a lake is called the thermocline.

2. _____ Water in a lake reaches its maximum density at four degrees Celsius.

3. _____ Lakes are totally self-contained ecosystems and are not influenced by terrestrial ecosystems surrounding them.

4. _____ Photosynthesis is greater than respiration at the compensation level of light.

5. _____ Algae and diatoms growing on the undersides of waterlily leaves are part of the periphyton.

6. _____ Macrophytes provide primary production for the limnetic zone of a lake.

7. _____ Macrophytes draw nitrogen and phosphorus from the open water of a lake.

8. _____ Eutrophic lakes are shallow and warm and have very little oxygen at their bottom.

9. _____ Hydrophytic plants include cattails and pondweeds.

10. _____ Red maple is a facultative wetland plant, growing both in dry uplands and forested wetlands.

11. _____ Wetlands are detrital systems.

12. _____ Hydrophytic vegetation will not grow in water-saturated soil.

13. _____ A moor is a blanket bog.

14. _____ Peatlands have low productivity.

15. _____ Hydroperiod includes the direction of water flow.

16. _____ The major source of nitrogen in peatlands is precipitation.

17. _____ Zonation about wetlands reflects the response of plants to hydroperiod.

18. _____ Macrophytes are the major energy source of headwater streams.

19. _____ Major energy source in riverine systems is FPOM.

20. _____ The river continuum is destroyed in regulated rivers.

21. _____ Pools of water behind dams take on the characteristics of a lake.

22. _____ Aufwuchs are a major source of primary production in headwater streams.

23. _____ Pools are the sites of primary production in streams.

24. _____ Spiraling is a mechanism for nutrient cycling in lakes and ponds.

Matching

Match the following zones of a pond or lake with the statements below.

A. Littoral zone; B. Limnetic zone; C. Profundal zone; D. Benthic zone.

1. _____ The dominant organisms found here are anaerobic bacteria.

2. _____ The main forms of life in this zone are phytoplankton and zooplankton.

3. _____ Cattails, sedges, and waterlilies grow in this zone.

4. _____ This zone is beyond the depth of effective light penetration.

5. _____ Its energy source is a rain of organic material produced in the layer above.

Match the following wetland types with the appropriate description below.

A. Shrub swamp B. Saline flats C. Swamp D. Bog E. Fen

6. _____ Waterlogged soil with a spongy covering of *Sphagnum* moss.

7. _____ Waterlogged soil with alder, willows, and buttonbush.

8. _____ A mire fed by water moving through mineral soil; dominated by sedges.

9. _____ A forested wetland.

10. _____ Inland saline area that floods after a rain.

Associate the following feeding groups in flowing water with their appropriate resource:

A. Piercers; B. Scrapers; C. Gougers; D. Shredders; E. Collectors

11. _____ Feed on leaves and other large organic particles.

12. _____ Burrow into waterlogged logs and limbs.

13. _____ Pick up FPOM drifting downstream.

14. _____ Feed on mosses and filamentous algae.

15. _____ Feed on algae coating on stones in streams.

Short Answer Questions

1. Why cannot wetlands be defined by vegetation alone?.

2. Contrast phosphorous cycling in a cattail marsh and tundra sphagnum bog.

3. How does hydroperiod influence the structure of a wetland?

4. If a lake appears to be a closed system, then why and how does it respond to the surrounding watershed?

ANSWERS

Key Term Review

1. limnetic
2. aufwuchs
3. oligotrophy
4. epilimniom, metalimnion, thermocline, hypolimnion
5. overturn
6. benthic zone
7. eutrophication
8. littoral zone
9. riverine wetlands
10. hydrophytic
11. lotic, lentic
12. trophogenic, tropholytic zones
13. benthic zone
14. phytoplankton
15. zooplankton
16. emergents
17. overturn
18. buffer systems
19. nekton
20. biological oxygen demand
21. cultural europhication
22. hypertrophic
23. dystrophic
24. marl lake
25. basin, watershed
26. shredders scrapers collectors
27. spiraling, uptake turnover
28. regulated
29. wetland
30. marsh, swamp, bog
31. basin wetlands fringe wetlands
32. riparian woods
33. vernal pools.
34. mires
35. blanket mires moors
36. fens
37. raised bogs quaking bogs
38. hydroperiod

Self Test

True and False

1. F	13. F
2. T	14. T
3. F	15. F
4. F	16. F
5. T	17. T
6. F	18. F
7. F	19. T
8. T	20. T
9. T	21. T
10. T	22. T
11. T	23. F
12. F	24. F

Matching

1. D	11. D
2. B	12. C
3. A	13. E
4. C	14. A
5. C	15. B
6. D	
7. A	
8. E	
9. C	
10. B	

Short Answer Questions

1. Wetlands cannot be defined by vegetation alone, because many wetland plants are facultative; they can grow in standing water as well as saturated soil, and even drier sites. Because of this growth pattern, vegetation shows a gradual transition from wetlands to upland sites. Other criteria must be considered. Aside from vegetation, a second important criterion is the presence of gley and organic soils. Another important criterion is the duration, frequency, depth, and season of flooding or the hydroperiod. These and other characteristics must be used to define a wetland.

2. In the spring new cattail shoots draw on the root reserves before they draw P from the soil. As the belowground pool shrinks, cattails draw P from the soil and accumulate P in aboveground biomass proportionately faster then they accumulate biomass. In late summer cattails remobilize P in roots but at a rate slower than they accumulate root biomass. Leaching and death of shoots return P drawn from deep soil onto the surface where it becomes available through decomposition. Because less P is returned than used, cattails must draw on a P supply from the soil.

In contrast, the P cycle in bogs is much tighter because there is little P to drawn from the substrate. *Rubus*, a typical trailing plant in bogs, increases its uptake of P from the roots to bud break. After that, as the temperature in the peat rises, the plant increases P in its stems, leaves, and roots. Late in summer the plant transfer much of the P to fruit, roots, and rhizomes. As the plant senescences in late summer, the plants lose P from its shoots and accumulate it in the roots and winter buds, where it is available for the next growing season.

3. The hydroperiod, the duration, depth, and frequency and season of flooding, can favor some plants and over others. Deep water that exists throughout droughts will favor submerged plants and deep water emergents. Periodic droughts that expose wetland bottoms stimulate germination of seeds of many emergent and annual aquatic plants. Brief flooding in spring will favor a wet meadow community. Thus, the character of any given wetland is influenced by the nature of its water regime.

4. A lake is hardly a closed system. Nutrients are carried by windblown leaves and other particulate matter. precipitation, inflowing streams and springs, groundwater seepage and other sources. Nutrients are carried away by outflowing streams , seepage through lake basin, evaporation, and other pathways.

Chapter 16
Saltwater Ecosystems

Learning Objectives

After completing this chapter you should be able to:
- Describe the major physical characteristic of the saltwater environment.
- Explain how tides result.
- Discuss the role of phytoplankton, zooplankton, nekton, and benthos in the open sea ecosystem.
- Describe the unique features of hydrothermal vents.
- Discuss the variations in productivity of the open sea and the food webs involved.
- Describe the zonation of rocky shore and adaptations of life to it.
- Point out the differences in life on sandy beaches and rocky shores.
- Discuss the energy base of the sand and mud shore ecosystems.
- Describe coral reefs and explain why they are of biological rather than geological origin.
- Define an estuary and discuss its major features.
- Explain the importance of the estuary to the coastal ecosystem.
- Describe the salt marsh and how tides and salinity influence its structure.
- Tell why salt marshes are highly productive.
- Identify some unique inhabitants of the salt marsh.
- Describe the unique features and importance of mangrove swamps.

Summary

After reading the chapter and before continuing to the material below, read the Chapter Summary on pages 357-358. You will find related material on the following pages: Coriolis effect, 91; ocean currents, 42-43; characteristics of water, 64-66; halophytes, 123-123; productivity of oceans, 172, 175-176; energy flow, 189-191; nitrogen cycle, 210-212; phosphorus cycle, 212-213; plant-herbivore interactions, 522-524; mutualism, 582-596; disturbance and community structure on rocky shores, 621-622.

Study Questions

1. What are the major regions of the oceans? (326)
2. What is the salinity of the open sea and how is it expressed? (327)
3. Compare the density of sea water with the density of fresh water, as the temperature of each changes. (328)
4. What is the halocline, thermocline, and pyncocline? (328)
5. How are waves generated? (328)
7. What causes tides? What are spring tides? Neap tides? (329-330)
8. What regions of the seas are richest in phytoplankton? Why? (311-332)
9. What is the food source of the nekton of the deep ocean? (332)
10. How are nekton organisms adapted for life in the deep? (332)
11. Distinguish between epifauna and infauna. (333)
12. Contrast the benthic substrate of the deep ocean with that of coastal waters? (333)
13. What feeding strategies are employed by bottom organisms? (333)
14. What are hydrothermal vents and what is unique about them? (333)
15. What factors influence the productivity of the open sea? (334-336)
16. Contrast the productivity of the open sea with that of coastal waters and explain the difference. (335)
17. What are the three major zones of the rocky shore and what are the outstanding features of each? (337-339)
18. How are barnacles and periwinkles adapted to withstand low tides? (340)
19. In what ways are tide pools microcosms of the sea? (340-341)
20. Contrast the energy source of a sandy shore with that of a rocky shore. (341)
21. What two basic groups occupy the sandy shore? (341-342)
22. How do tidal shores relate to coastal and marine ecosystems? (343-344)
23. What are the outstanding features of the coral reef? 344-345)
24. What is an estuary? (345)
25. Discuss the vertical and horizontal stratification of salinity in the estuary? (346)
26. What is tidal overmixing and what influences govern it? (346)
27. How is an estuary a nutrient trap? (347)
28. What is the interrelationship between marine fish and an estuary? Oyster reefs and an estuary? (347-348)
29. What two factors make the estuary an ideal nursery for marine fish? (347)
30. What are salt marshes? Where are they located and how do they develop? (348-349)
31. What are the three dominant grasses of a salt marsh? What features determine their distribution within the marsh? (350-352)
32. In what way is *Spartina alterniflora* adapted to the salt water environment? (351)

33. What are the dominant animals in the salt marsh and how do they relate to tidal regimes? (352-353)?
34. What is the major type of food chain in the salt marsh? (353)
35. Why are the salt marsh one of the most productive ecosystems? (354)
36. What is the role of sulfur in the energy flow of salt marshes? (354)
37. Briefly describe the nitrogen cycling in a salt marsh. (355)
38. What are the major features of a mangrove swamp? (355)
39. How do mangrove swamps respond to tidal flooding? (355-356)
40. What is the ecological and commercial importance of mangrove swamps? (356)
41. What is the role of hydroperiod in the nutrient cycle of mangrove swamps? (356-357)

Key Terms and Phrases

pelagic region	sublittoral zone	epiflora
benthic region	littoral zone	epifauna
photic zone	halocline	infauna
mesopelagic layer	pycnocline	meiofauna
bathypelagic layer	Langmuir cells	fringing reef
neritic province	Ekman spiral	barrier reef
oceanic province	spring tide	atoll
benthos	neap tide	zooxanthella
benthipelagic province	hydrothermal vent	estuary
hadalpelagic province	microbial loop	salt pan
bathyal zone	supralittoral fringe	mangals
hadal zone	infralittoral fringe	salt marsh
index of salinity	upwelling	nekton
bioluminesence	abyssal zone	sublittoral fringe

Key Term Review

1. The open body of water that makes up the ocean is the _____ _____..

2. The amount of chlorine in sea water is used as an _____ ____ _____.

3. When surface water is replaced by water moving upward from the deep, the process is called _____.

4. The bottom region of the ocean is the _____ _____.

5. Plants and animals that live on the rocky bottoms of the ocean make up the _____ and _____.

6. _____ borrow into the substrate of the ocean floor and sandy beaches.

7. The _____ _____, a subdivision of the pelagic zone ranging down to 200 meters, is characterized by sharp gradients in light, temperature, and salinity.

8. In the _____ _____the temperature change is gradual with little seasonal variation.

9. The water that overlies the continental shelf is the _____ _____; the water that lies beyond the continental shelf is the _____ _____.

10. The _____ _____ lies over the major plain of the ocean to a depth of 6000 meters; waters of the deep oceanic trenches are the _____ _____.

11. On the sea bottom, the benthic region, the _____ _____ covers the continental shelves down to 4000 meters; ,the _____ _____ goes down to 6000 meters; the benthic zone of the deep ocean trenches is the _____ _____.

12. The benthic zone underlying the neritic pelagic zone on the continental zone is the _____ _____; the intertidal zone is the _____ _____.

13. The vertical gradient of salinity in the ocean is called the _____; the zone of rapid change in density is the _____.

14. When the Earth, moon, and sun are in line with each other, the gravitational pull of the sun and moon result in very high tides known as _____ _____. When the gravitational pull of the sun and moon interferes with each other a _____ _____ results.

15. Vertical cells of spinning water just below the surface generated by wind are called _____ _____.

16. The phenomenon of successive layers of ocean water set into spiral motion by the energy of the wind blowing on the surface, until the deepest layer is moving counter to the flow on the surface is called the _____ _____.

17. Life on the ocean bottom makes up the _____; free-swimming organisms make up the _____.

18. Organisms of the bathypelagic region rely on _____ to attract food and recognize their species.

19. High temperature deep-sea springs that support a rich diversity of endemic deep-sea life are _____ _____.

20. The base of the oceanic food web consists of heterotrophic bacteria and nanoflagellates that take up dissolved organic matter produced by plankton. The consumption of bacteria by nanoplankton produces a feeding structure called the _____ _____.

21. The transition zone from land to sea is the _____ _____.

22. The area of shore covered and uncovered by daily tides is the _____ _____; that part of the shore uncovered only by the spring tides is the _____ _____.

23. Copepods, ostracods, and nematodes living within the sand and mud of the shore make up the _____.

24. Coral reefs are of three types: _____ _____ form parallel to shorelines and are separated from land by lagoons; _____ _____ project directly seaward from the shoreline ; _____ are coral islands.

25. Algae living in corals are _____.

26. An _____ is an ecosystems where fresh water meets and mixes with salt water.

27. _____ replace salt marshes in tropical regions.

28. Elliptical depressions and flats in the high marsh are _____ _____.

29. A flat, grass-covered coastal area within an estuarine ecosystem in a temperate climate is called a _____ _____.

Self Test

True and False

1. _____ Sea water reaches its greatest density at 4°C.

2. _____ The productivity of the open ocean is high.

3. _____ The highest concentration of phytoplankton is in the photic zone.

4. _____ Organisms that possess bioluminescence live in the pelagic zone.

5. _____ The most abundant element in the ocean is sodium.

6. _____ Baleen whales make up part of the nekton.

7. _____ Seawater reaches its freezing point slightly below that of pure water.

8. _____ Neap tides are most likely to occur when the moon is full.

9. _____ The largest part of the energy flow in pelagic ecosystems depends upon bacteria and protists.

10. _____ Water that overlies the continental shelf is part of the neritic zone.

11. _____ The mesopelagic zone is an area that shows a sharp seasonal variation in temperature and light.

12. _____ Zooplankton are the major herbivores in a pelagic ecosystem.

13. _____ The distribution and composition of phytoplankton is influenced by light, temperature, and nutrients.

14. _____ The giant clam, *Calyptogena magnifica,* inhabiting hydrothermal vents has a symbiotic relationship with chemosynthetic bacteria.

15. _____ Tropical seas have a thermal stratification that varies with the seasons.

16. _____ Burrowing worms and clams living in the mud and in the floor of the ocean are part of the epifauna.

17. _____ Sediments below 6000 meters in the ocean contain little organic matter, but do contain large quantities of silica an aluminum oxides.

18. _____ Chemosynthetic bacteria near hydrothermal vents in the ocean act as primary producers oxidizing reduced sulfur compounds.

19. _____ A major energy base for marine food webs is heterotrophic bacteria.

20. _____ Oceanic zooplankton move up to the surface waters at night to feed and descend to deeper water by day.

21. _____ The energy base for the sandy beach is organic matter deposited by tides.

22. _____ The size of sand particles has little influence on the nature of life in a beach.

23. _____ Mussels are highly resistant to desiccation.

24. _____ Where wave action is heavy on rocky shores, periwinkles are abundant.

25. _____ Tide pools show little daily variation in salinity and temperature.

26. _____ Beach hoppers and ghost crabs inhabit the supralittoral zone of the sandy beach.

27. _____ Sandy beaches are very dependent on organic matter produced away from the beach.

28. _____ Organisms living within the mud and sand of a beach must tolerate wide fluctuations in temperature and salinity.

29. _____ Predation and wave action produce the patchiness in the distribution of life on rocky shores.

30. _____ Barnacles prevent water loss at low tide by secreting mucus.

31. _____ Chemosynthetic bacteria are decomposers of organic matter.

32. _____ Most of the organisms living in an estuary are benthic.

33. _____ In an estuary at the mouth of a river, fresh water from the river sits below sea water.

34. _____ The salt marsh is a detrital system.

35. _____ *Spartina alterniflora* can tolerate being flooded only for a few hours each day.

36. _____ Deltas develop at the mouths of rivers.

37. _____ Salinity in an estuary remains stable and constant, showing no seasonal or daily fluctuations.

38. _____ Organisms living in estuaries are essentially marine and must be able to tolerate seawater.

39. _____ The salt marsh is one of the most productive habitats in the marine environment.

40. _____ The highest rate of productivity in mangrove forests occurs in those forests that experience daily tides.

41. _____ Mangrove trees have leaves equipped with salt pores that help maintain a proper salt balance in them.

Matching

Associate each of the following regions of the ocean with their appropriate descriptions:

A. Photic zone   C. Bathypelagic zone   E. Neritic zone
B. Mesopelagic zone   D. Benthic zone   F. Hadal zone

1. _____ The ocean bottom.

2. _____ Deepest part of the ocean.

3. _____ Water over continental shelves.

4. _____ Complete darkness is the order here, except for some bioluminescence.

5. _____ Water near the surface to 200 meters with sharp gradients in light intensity, temperature, and salinity.

129

6. _____ Contains the oxygen-minimum layer and high concentrations of phosphate and nitrate.

Associate each of the following intertidal zones with zones with the statements below:

A. Supralittoral zone   B. Littoral zone   C. Infralittoral zone

7. _____ Home of kelp and bladder rockweed.

8. _____ Often called spray zone on rocky shores because water rarely covers this area except during high spring tides.

9. _____ Area is dominated by barnacles.

10. _____ Starfish and sand dollars inhabit this zone on sandy shores.

11. _____ Under water most of the time, becoming exposed only at extreme low tide. Area subject to violent wave and current action.

12. _____ Ghost crabs run over this zone by night and live in burrows in the sand by day.

13. _____ Lugworms and clams lie hidden in the sand in this zone.

14. _____ Area covered and uncovered by daily tides; the upper region of this zone on rocky shores would be dominated by periwinkles.

Associate each of the following zones of a salt marsh with the descriptive statement:

A. Low marsh   C Tidal creeks   E. Salt pans
B. High marsh   D. Salt meadow

15. _____ Mud algae, diatoms, marsh periwinkle, outflow of fresh water.

16. _____ Anaerobic mud, marsh periwinkle, *Spartina alterniflora*.

17. _____ Floods a few hours each day, short form of *Spartina alterniflora*, low tidal exchange, open canopy.

18. _____ *Spartina patens*, mounds of detritus, marsh snails and mice are common inhabitants.

19. _____ Algal crusts, crystallized salt, edges support *Salicornia* and *Distichlis*.

Short Answer Questions

1. In the dark world of the benthic regions there is an amazing diversity of life. What is the major energy source of these organisms? What feeding strategies do they employ?

2. What is the role of bacteria in sand and mud shores?

3. Patchy distribution is common among rocky intertidal organisms. What physical and biological interactions give rise to this patchiness?

4. What is the importance of zooxanthellae to the coral colony? (refer to pages 581-582, Figure 27.5)

5. What are some of the special adaptations of *Spartina alternifolia* to the salinity of the salt?

6. How does the reproductive biology of mangroves enable them to colonize shallow water areas of the tropics?

ANSWERS

Key Term Review

1. pelagic region
2. index of salinity
3. upwelling
4. benthic region
5. epiflora, epifauna
6. meiofauna
7. photic zone
8. mesopelagic layer
9. neritic province, oceanic province
10. benthipelagic province hadalpelagic province
11. bathyal zone, abyssal zone hadal zone

12. sublittoral zone littoral zone
13. halocline, pycnocline
14. spring tide, neap tide
15. Langmuir cells
16. Eckman spiral
17. benthos, nekton
18. bioluminescence
19. hydrothermal vents
20. microbial loop

21. supralittoral fringe
22. infralittoral fringe sublittoral fringe
23. infauna
24. barrier reefs fringing reefs atolls
25 zooxanthella
26. estuary
27. mangals
28. salt pans
29 salt marsh

Self Test

True and False

1. F	15. F	29. T	
2. F	16. F	30. F	
3. T	17. T	31. F	
4. F	18. T	32. T	
5. F	19. T	33. F	
6. T	20. T	34. T	
7. T	21. T	35. F	
8. F	22. F	36. T	
9. T	23. T	37. F	
10. T	24. F	38. T	
11. F	25. F	39. T	
12. T	26. T	40. T	
13. T	27. T	41. T	
14. T	28. F		

Matching

1. D	14. B
2. F	15. C
3. E	16. A
4. C	17. B
5. A	18. D
6. B	19. E
7. B	
8. A	
9. B	
10. A	
11. C	
12. A	
13. B	

Short Answer Questions

1. Benthic food webs depend upon the rain of detrital material from above. Benthic organisms may a) filter suspended material from the water, as do stalked coelenterates; b) collect particles that settle on the surface sediments, as sea cucumbers do; c) be selective or unselective deposit feeders, such as polychaetes; or d) they may be predators, such as brittlestars.

2. Bacteria in mud and sand shores use up oxygen, creating anoxic conditions and break down organic compounds for use by other organisms. They form the basis of certain food webs. Chemosynthetic bacteria are primary producers on the shore.

3. Patchiness results from both abiotic and biotic interactions. Wave action clears life from patches of rocks, opening the area for colonization by settling larvae. These larvae may select different areas on the rocks. Some species compete for space; others are territorial. Social behavior may lead to clustering of individuals in crevices on rocks. Grazing by limpets eliminates algal growth on rocks. Predation by large predators, such as *Piaster,* reduces competition among species allowing greater diversity.

4. Zooxanthellae, living within the coral, produce food through photosynthesis and transfer some of this to the coral. Though the photosynthetic process, they produce oxygen used by coral. They use metabolic wastes, especially nitrates and phosphates, produced by the coral colony. These activities increase the rate of calcification of the coral colonies.

5. *Spartina alternifolia* is well adapted to live in a water-logged saline environment. It selectively concentrates sodium chloride at a level higher in its cells than in surrounding seawater, maintaining its osmotic integrity. It rids itself of excessive salts through salt-secreting cells in its leaves. To get air to its roots in anoxic mud, *Spartina* has hollow air-carrying tubes leading from leaves to roots.

6. Mangrove seeds germinate on the tree. The weight of the seed is concentrated on the bottom. When the seeds drop they float upright. When they strike bottom, they can put out roots directly into the sediments and send out leaves.

PART 5

POPULATION ECOLOGY

CHAPTER 17
PROPERTIES OF POPULATIONS

Chapter Outline

Defining Populations
  Unitary and Modular Populations
  Populations as Genetic Units
  Populations as Demes
Density and Dispersion
  Crude vs Ecological Density
  Patterns of Dispersion
Age Structure
  Age Structure in Animals
  Age Structure in Plants

Sex Ratios
Mortality and Natality
  The Life Table
  Survivorship and mortality Curves
  Animal Natality
  Plant Natality
  Fecundity Curves and Reproductive Values
Summary

Learning Objectives

After completing this chapter you should be able to:
- Distinguish between modular and unitary populations.
- Distinguish between a genet and a ramet.
- Describe a population as a genetic unit and as a deme.
- Distinguish between crude and ecological density.
- Explain the differences between random, uniform, and clumped distribution.
- Distinguish between immigration, emigration, and migration.
- Describe the age structure of a population.
- Distinguish between stable age distribution and stationary age distribution.
- Explain the significance of age structure to a population.
- Describe the significance of sex ratio in populations.
- Define death rate, mortality rate, probability of dying, and life expectancy.
- Explain the structure of a life table.
- Describe and tell the differences between survivorship curves and mortality curves.
- Define natality, crude birth rate, specific birth rate, and net reproductive rate.
- Describe a fecundity table and a fecundity curve.
- Define reproductive value and its explain its significance.

Summary

After reading this chapter and before continuing with the material below, read the Chapter Summary on page 387. You will find related material on the following pages; adaptation, 30; population growth, 392-396; dispersal, 413-417; mating systems, 428-430; reproductive costs, 436-446; Appendix B, 729-732.

Study Questions

1. What is a population? (362)
2. What distinguishes a modular population from a unitary population? Give an example of each. (362-363)
3. Distinguish between a genet, a ramet, and a clone. How do they relate to a modular population? (363-364)
4. What is a gene pool? A deme? (364)
5. What is the difference between ecological effective distance and ecological effective density? (364)
6. What is population density? Crude density? Ecological density? (365)
7. What are the three basic types of spatial distribution of populations? Which one is most common and why? (366-368)
8. What is the meaning of grain to local distribution of populations? (368)
9. How are populations dispersed in time? (368-370)
10. What are three basic types of dispersal movements? (368-371)
11. What is the significance of age structure in a population? (372-374)
12. What is an age pyramid? What does it tell us about a population? (372-373)
13. What is the difference between a stationary age distribution and a stable age distribution? (372)
14. What are some of the problems of and the usefulness of applying the concept of age structure to a plant population? (373-374)
15. What is the significance of sex ratios to a population? (374-375)
16. Distinguish between death rate, mortality rate, probability of dying and life expectancy. (376)
17. What is natality, crude birth rate, and specific birth rate? (376)
18. What is a life table? What are the differences between a dynamic life table, a time-specific life table, and a dynamic-composite life table? (376-377)
19. What are the major columns of a life table? Explain each. (376-377, 729-730)
20. What is a survivorship curve? What are the three basic types? (380-382)
21. What is a mortality curve? How does it differ from a survivorship curve? (362)
22. What is the net reproductive rate? What are some problems in determining net reproductive rate to plants? (383-384)
23. What is a fecundity table? (385-386; 730-731)
24. What is the difference between gross reproductive rate and net reproductive rate? (386, 730-731)
25. What is reproductive value and how is it calculated? (386, 731-732)

Key Terms and Phrases

population	coarse grain	natality
demography	range	physiological natality
module	emigration	realized natality
genet	immigration	crude birthrate
ramet	migration	age specific birthrate
clone	stable age distribution	cohort
gene pool	stationary age distribution	horizontal life table
deme	primary sex ratio	dynamic life table
ecological effective distance	secondary sex ratio	time specific life table
ecological neighborhood	death rate	survivorship curve

ecological effective density
density
crude density
ecological density
grain
dynamic-composite life table

crude death rate
probability of dying
probability of surviving
life expectancy
fine grain
clumped distribution

mortality curve
net reproductive rate
fecundity curve
reproductive value
uniform distribution
random distribution

Key Term Review

1. A group of interbreeding or potentially interbreeding populations occupying a given area at a given time is a _____.

2. A single plant with its own genetic characteristics is termed a _____; subpopulations of that plant formed by roots and suckers are called _____, also popularly known as _____.

3. A population of 100 rats per city block is a measure of _____ _____.

4. Each stem or root of a plant that is part of a larger unit is called a _____.

5. The study of vital statistics of populations is _____.

6. A small local population of a species is called a _____.

7. The sum of all genetic information carried by a population it is _____ _____.

8. When a distribution of each individual in a population is independent of other individuals, it is called _____.

9. When individuals within a population of violets are clustered about wet places in a woods. they are exhibiting a _____ _____.

10. Christmas trees growing in row in a plantation are a good example of _____ _____.

11. Regional distribution of a population make up its _____.

12. Annually elk come down out of the mountains for winter and return to the high mountain meadows in summer. Their movements would be a _____. Some individual elk move out the area permanently to occupy new habitat. This movement represents _____. Their movement into the new habitat would be _____.

13. Sex ratio at conception is the _____ _____ _____; the sex ratio at birth or hatching is the _____ sex ratio.

14. If a population has a fixed proportion of individuals in its age classes, it exhibits a _____ _____ _____.

139

15. A _____ _____ _____ requires that birth rates and death rates must remain the same.

16. The death of 30 individuals per 1000 is a measure of _____ _____ _____.

17. The number of deaths during a given time interval divided by the average population size during that period is the _____ _____.

18. The number of individuals dying in a given time interval divided by the number of individuals alive at the beginning of the time interval gives the _____ __ _____. Its complement is the _____ _____ _____.

19. The average number of years to be lived into the future by members of a given age in a population is the _____ _____.

20. The birth of new individuals in a population is _____. The maximum number of births under ideal conditions is _____ _____.

21. The number of successful births that actually occurs over a given period of time is _____ _____.

22. The number of births per 1000 individuals is _____ _____.

23. A more accurate measure of births is the number of offspring produced per unit time by females in different age classes, the _____ _____.

24. A _____ _____ provides a schedule of mortality and survival of a standard number of individuals all born in the same time interval, called _____.

25. A _____ _____ _____ is based on a group of individuals all born within a given time frame and followed through to the death of the last individual.

26. A _____ _____ life table is based on a sample of a population and aging the organisms to obtain a distribution of age classes during a single time period.

27. A _____ - _____ _____ _____ considers a cohort a composition of a number of animals marked over a period of time rather than just one birth period.

28. The number of organisms living within a given area of acceptable habitat is _____ density.

29. Individuals spaced so that each does not affect the other is the _____ _____ _____. The _____ _____ density, the highest density a population an attain without one individual crowding another.

30. The pattern of local distribution is called _____. A pattern in which an individual is likely to have another member of its own species as its neighbor is called _____ _____. If the likelihood is that an individual will have as its neighbor an individual of another species the pattern is called _____. _____.

31. A _____ _____ plots the age specific mortality in terms of remaining living individuals through time, whereas a _____ _____ plots the ratio of number dying during a time period to the number alive at the beginning of the period.

32. The number of female offspring left during a lifetime by a female is the _____ _____ _____.

33. A _____ _____ depicts the number of offspring per individual per age class.

34. The average contribution to the next generation that individuals of a given age group give to the next generation between their current age and death is termed their _____ _____.

Self Tests

True and False

1. _____ A survivorship curve that is strongly convex is typical of fish.

2. _____ A plot of the proportion of a cohort of organisms still alive from birth through time is a mortality curve.

3. _____ The age distribution of a population is the proportion of individuals in each age group; in a changing environment, the distribution is seldom maintained in a stable condition for very long.

4. _____ The letter J best describes the shape taken by the mortality curve of most mammals.

5. _____ If the position of each individual in a population is independent of each other, the distribution is said to be clumped.

6. _____ A survivorship curve that is more or less linear is typical of birds.

7. _____ Emigration is a permanent movement out of an area.

8. _____ A time-specific life table records the fate of a group of animals all born at the same time.

9. _____ Climatic changes can influence a species' range.

10. _____ Territoriality in animals will result in a clumped pattern of distribution.

11. _____ A single plant is a collection of subpopulations.

12. _____ Organisms are distributed in time as well as in space.

13. _____ The age structure of a population has little influence on the mortality and natality rates.

14. _____ Most natural populations reach a stable age distribution after a few years.

15. _____ A mortality curve derived from a life table is plotted from the d_x column.

16. _____ The secondary sex ratio is weighted toward males.

17. _____ Plant demographers have two levels of mortality to consider, that of genets and of metapopulations.

18. _____ If mortality rates are constant for all age groups in a population, the survivorship curve will be concave.

19. _____ Germination in plants is equivalent to birth in animals.

20. _____ Type I survivorship curves are typical of populations in which the mortality rate of juveniles is very high.

Matching

Match the life table categories on the right with the statement on the left::

1 _____ Data for the fecundity curve.   A. l_x

2. _____ Mortality.   B. d_x

3. _____ Life expectancy.   C. q_x

4. _____ R, net reproductive rate.   D. e_x

5. _____ Data for mortality curve.   E. m_x

6. _____ Survivorship.   F. $\Sigma\, l_x m_x$

142

Life Table Questions

The following is a life table of a hypothetical population of species X.

x	l_x	d_x	q_x	e_x	m_x	$l_x m_x$
0-1	1.00	0.45	0.45	1.8	0.0	0.00
1-2	0.55	0.20	0.36	1.2	0.5	0.28
2-3	0.35	0.20	0.57	1.1	1.0	0.35
3-4	0.15	0.10	0.66	0.9	2.0	0.30
4-5	0.05	0.04	0.80	0.7	1.0	0.05
5-6	0.01	0.10	1.00	0.5	0.0	0.00

Answer the following questions as they relate to the above life table:

1. _____ At age 3 (x = 3-4) what is the probability of dying?

2. _____ At age 4 (x = 4-5) what is the life expectancy?

3. _____ How many individuals died in age class 0 (x = 0-1)? Multiply the value by 1000 for the answer.

4. _____ What is the fecundity for age class 2 (x = 2-3)?

5. _____ What is the net reproductive rate? (Calculate it.)

6. Draw a survivorship curve and a mortality curve based on the data in the above life table, labeling both axes.

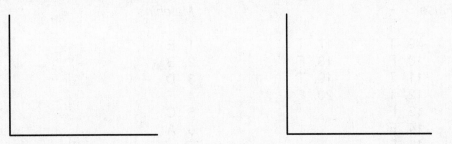

Short Answer Questions

1 What is the difference between crude density and ecological density? Why is it important to have some measure of ecological density?

2. What information about a population can be gained from age pyramids? Why are age pyramids more historical than predictive?

ANSWERS

Key Term Review

1. population
2. genet, ramet, clone
3. crude density
4. module
5. demography
6. deme
7. gene pool
8. random
9. clumped distribution
10. uniform
11. range
12. migration, emigration, immigration
13. primary, secondary
14. stationary age distribution
15. stable age distribution
16. crude death rate
17. death rate
18. probability of dying
    probability of survival
19. life expectancy
20. natality, physiological natality
21. realized natality
22. crude birthrate
23. age-specific birthrate
24. life table, cohort
25. dynamic life table
26. time-specific life table
27. dynamic-composite life table
28. ecological density
29. ecological effective distance
    ecological effective density
30. grain, coarse grain. fine grain
31. survivorship curve, mortality curve
32. net reproductive rate
33. fecundity curve
34. reproductive value

Self Test

True and False

1. F	9. T	17. T
2. F	10. F	18. F
3. T	11. T	19. T
4. T	12. T	20. F
5. F	13. F	
6. T	14. F	
7. F	15. F	
8. T	16. T	

Matching

1. E
2. B
3. D
4. F
5. C
6. A

Life Table Questions:

1. 0.36   2. 0.7   3. 450   4. 1.0   5. 0.98

6.

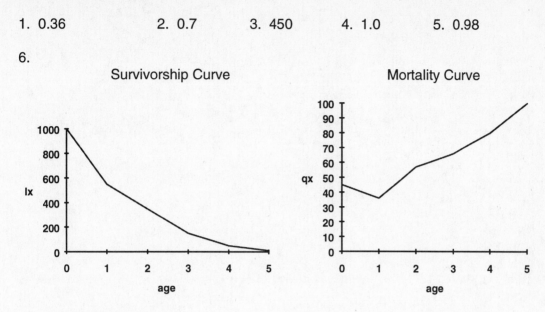

Short Answer Questions

1. Crude density is simply a measure of the number of organisms in a given area without any reference to the distribution of those organisms. Ecological density relates the number of organisms per given area of actual living space or habitat. Since the area of suitable living space may be much less within the given area, the organisms, restricted to that area, may be more crowded than crude density implies. For example, there may be a certain number of fish per mile of stream, implying that the fish are uniformly distributed throughout. However, these fish may be restricted to pools within the stream. The ecological density, the number of fish per pool, would be much higher in a more restricted area of the stream. Information on ecological density is useful in managing or providing habitat for a particular species and maintaining its population.

2. Age pyramids are a graphic portrayal of the relative proportion of the various age classes in a population. They depict, in particular, the ratio of prereproductive, reproducive, and postreprodctive age classes. Age pyramids are historical. A sequence of age pyramids tells the trends of various age classes through time and provides insights into a population's past history. Because future environmental and population changes cannot be predicted, age pyramids cannot, except in a very rough way, predict future population growth.

CHAPTER 18
POPULATION GROWTH AND REGULATION

Chapter Outline

Rate of Increase
Population Growth
   Exponential Growth
   Logistic Growth
   Time Lags
Density-Dependent Regulation
   Intraspecific Competition
   Growth and Fecundity
   Plant Biomass

Density-Independent Influences
Population Fluctuations and Cycles
Key Factor Analysis
Extinction
Summary

Learning Objectives

After completing this chapter you should be able to:

- Discuss population increase in terms of lambda, net reproductive rate, and rate of increase.
- Explain the difference between exponential and logistic population growth.
- Discuss the role of time lags in population growth.
- Describe the difference between density-dependent and density-independent influences on population growth.
- Define intraspecific competition and distinguish between scramble and contest types of competition.
- Discuss the effects of population density on growth and fecundity of individuals.
- Distinguish between population fluctuations and population cycles.
- Tell how key factor analysis can be used to detect density-dependent influences in a population.
- Discuss population extinction and the difference between demographic extinction and stochastic extinction.

Summary

After reading this chapter and before continuing with the material below, read the Chapter Summary on page 409. You will find related material on the following pages: plant responses to moisture, 70-75; animal responses to moisture, 75-79; genetic drift, 471-472; viable populations, 475-476; plant-herbivore interactions, 522-524; predator-prey cycles, 543-545; habitat fragmentation, 614-618.

Study Questions

1. Define the rate of increase. What is the difference between lambda λ, R_o, and r? What is the relationship among them? (390-392)
2. Distinguish between exponential growth and logistic growth. Give the formula for each. (392-395)
3. Define carrying capacity or K. (393)
4. At what point on the logistic curve is growth maximal? (395)
5. What are the weaknesses and limitations of the logistic growth equation? (395)
6. Why do populations fluctuate about some theoretical value of K? (395)
7. What time lags have to be built into the equation? (395-396)
8. Distinguish between density-dependent regulation and density-independent influences on population growth. (396, 402)
9. Define competition. What is scramble competition? Contest competition? (397-398)
10. What is the relationship among population density, biomass growth, and fecundity? (397-402)
11. What is the -3/2 power law of self-thinning? How does it relate to plant biomass accumulation and plant density? (400-402)
12. What are some examples of density-independent influences on populations? Why can't density-independent influences regulate a population? (402)
13. How can density-independent influences affect density-dependent mechanisms? (402)
14. What is key factor analysis? How can it be used to gain insights into density-dependent influences in a population? (406-407)
15. Distinguish between a population fluctuation and a population cycle. (403-406)
16. Why do species go extinct? (407-408)
17. What is meant by demographic stochasticity and environmental stochasticity as they relate to the extinction process? (408-409)

Key Terms and Phrases

age distribution
mean cohort generation time
intrinsic rate of natural increase
iteration
doubling time
exponential growth
logistic growth
carrying capacity
inflection point
time lag
reaction time lag
reproductive time lag
stable limit cycle
density-dependent

scramble competition
exploitative competition
contest competition
interference competition
self-thinning
law of constant final yield
-3/2 power law
population cycle
population fluctuation
resilience
key factor analysis
deterministic extinction
stochastic extinction
density-independent

Key Term Review

1. A cold wet spring that caused the death of young turkey poults would have a
    _____ effect on the turkey population.

2. _____ _____ occurs when a population colonizes a new habitat.

3. The _____ _____ of a population is determined by its limiting resources.

4. _____ _____ occurs when there is some limit placed on population growth by the environment.

5. _____ _____ occurs when all the individuals within a population share a resource equally. It also is called _____ _____.

6. The proportion of individuals in the various age classes in a population is the population's _____ _____.

7. The measure of the instantaneous rate of change of a population size per individual is the _____ _____ ____ _____ _____.

8. The _____ _____ of a population results when $N_t / N_o = 2$.

9. The mean period of time elapsing between the birth of the parent and the birth of its offspring is the _____ _____ _____ _____.

10. _____ _____ _____ is based on a biological or environmental condition associated with mortality that causes major fluctuations in population size.

11. The point on the logistic curve where population growth is maximal is the
    _____ _____.

12. Time lags are involved in population recovery. A lag between environmental change and a corresponding change in population growth is a _____ _____ _____. A lag between environmental change and a change in the length of gestation is a _____ _____ _____.

13. Competition between individuals of the same species for a resource in short supply is _____ _____.

14. If one or more individuals within a population establishes a claim to an environmental resource _____ _____ or _____ _____ results.

15. _____ _____ comes about in a population through some force or change from which there is no escape. _____ _____ comes about from normal random changes in local populations.

149

16. How fast a population declines from above and how quickly it increases from below some equilibrium point is a measure of its _____.

17. _____ _____ results from time lags in birthrates and death rates in local populations.

18. Population changes more regular than would be expected by chance are _____ _____.

19. The progressive decline in density in a plant population of growing individuals is known as _____.

20. Plant ecologists have observed that the final yield in the plant biomass of a species from all starting densities becomes constant. This phenomenon is known as the _____ ____ _____ _____._____.

21. The slope of the logarithm of increasing plant weight plotted against the logarithm of decreasing density of survivors gives rise to the _____ _____ _____.

22. When a population fluctuates about some equilibrium level, with a certain period and amplitude, it exhibits a _____ _____ _____.

23. The rapid spread of disease through a population is a _____ effect.

Self Test

True and False

1. _____ A zero rate of population growth results when the death rate is less than the birthrate.

2. _____ Habitat destruction is a major cause of species extinction.

3. _____ No environment can support sustained exponential growth.

4. _____ The rate of increase of a population is constant, unaffected by environmental alterations.

5. _____ Influences on population growth by the purely physical environment are density independent.

6. _____ Density-dependent effects increase in intensity as a population grows larger.

7. _____ The maximum population size an area can support is dependent upon the rate of increase.

8. _____ The carrying capacity of an ecosystem is determined by the availability of space, water, light, and nutrients.

9. _____ The demise of the dinosaurs is an example of stochastic extinction.

10. _____ Weather is a density-independent factor that can affect population growth.

11. _____ Sparse populations often face increased death rates because predation pressure on the remaining individuals is high.

12. _____ Competition for resources is a density-independent influence.

13. _____ Contest competition results in some individuals in a population getting more resources than others.

14. _____ Density-dependent population regulation influences a population in proportion to its size.

15. _____ Logistic growth occurs when populations are not crowded.

16. _____ Competition for environmental resources may act as a negative feedback and slow population growth.

17. _____ The lynx and the snowshoe hare show population oscillations of approximately three to four years.

18. _____ Population extinction has been spread rather evenly across Earth's geological history.

19. _____ The differential equation for exponential growth states that the rate of increase is directly proportional to population size.

20. _____ The growth of populations results from negative feedback.

21. _____ A contest type of competition results in a wastage of resources.

22. _____ Exponential growth is independent of population density.

23. _____ Regardless of their starting densities, plant populations will converge on some common density that will decrease with time.

24. _____ All plants conform to the -3/2 power law.

25, _____ Carrying capacity is invariable, never changing through time.

26. _____ Populations of small-bodied animals fluctuate more widely than do populations of large-bodied animals.

Matching

Match the following symbols or formulas on the right with the appropriate definition on the left:

1. _____ Net reproductive rate.

   A. 0.693/r

2. _____ Finite multiplication rate.

   B. $dN/dt = rN$

3. _____ Doubling time.

   C. r

4. _____ Exponential growth.

   D. $dN/dt + rN(K - N/K)$

5. _____ Intrinsic rate of increase.

   E. $\Sigma l_x m_x$

6. _____ Logistic growth.

   F. λ

Short Answer Questions

1. Distinguish among net reproductive rate, finite rate increase, and annual rate of increase.

2. How do increasing population densities affect the growth and fecundity of individuals in the population?

3. Why do populations of small animals fluctuate more widely than those of large bodied animals?

ANSWERS

Key Term Review

1. density-independent
2. exponential growth
3. carrying capacity
4. logistic growth
5. scramble competition
   exploitative competition
6. age distribution
7. intrinsic rate of natural increase
8. doubling time
9. mean cohort generation time
10. key factor analysis
11. inflection point
12. reaction time lag
    reproductive time lag
13. intraspecific competition
14. contest competition
    interference competition
15. stochastic extinction
    deterministic extinction
16. resilience
17. population fluctuation
18. population cycle
19. self-thinning
20. law of constant final yield
21. -3/2 power law
22. stable limit cycle
23. density-dependent

Self Test

True and False

1. F	10. T	19. T
2. T	11. T	20. F
3. T	12. F	21. F
4. F	13. T	22. T
5. T	14. T	23. T
6. T	15. F	24. F
7. F	16. T	25. T
8. T	17. F	26. T
9. T	18. F	

Matching

1. E
2. F
3. A
4. B
5. C
6. D

Short -Answer Questions

1. The net reproductive rate, R, is derived from life table survivorship and fecundity data. It represents the number of offspring (usually female) left during a lifetime by an newborn individual (usually female). The finite rate of increase, λ, is based on a change in population size from one period of time, usually a year, to another. The intrinsic rate of natural increase, r, is the instantaneous rate of change in a population size per individual. It is derived from R, net reproductive rate, and can be used to estimate λ.

153

2. As populations increase they reach a point where resources are insufficient to meet the resource needs of individuals. As competition for a limited supply of food, in particular, increases, individuals may receive enough for body maintenance, but not enough for growth. As growth is stunted, so is the ability for the individual to produce young. because of delayed sexual maturity, failure to conceive, or inability to nourish any young produced.

3. Small-bodied animal populations fluctuate more widely than large-bodied animal populations because small animals have shorter lives, die more quickly, and thus decrease more dramatically. However, small-bodied animals also reproduce more rapidly and more often, enabling them to recover from their loses in a short period of time. Because they live longer, large-bodied animals possess more stability in their populations. However, they cannot recover rapidly from population losses because of lower reproductive rates and need a longer timer to reach sexual maturity.

CHAPTER 19
INTRASPECIFIC COMPETITION

Chapter Outline

Density and Stress
Dispersal
  Who Disperses?
  Why Disperse?
  Does Dispersal Regulate Populations?

Social Interaction
  Social Dominance
  Territoriality
  Home Range
Summary

Learning Objectives

After completing this chapter you should be able to:
- Discuss the role of physiological stress in dense populations.
- Describe the role and nature of dispersal in animal populations.
- Distinguish between different types of dispersal.
- Explain what is meant by a population sink.
- Distinguish between social dominance and territoriality and explain their possible roles in population regulation.
- Discuss the types of territories and territorial defense.
- Speculate on the reasons why some animals defend territories.
- Distinguish between a home range and a territory.

Summary

After reading this chapter and before continuing with the material below, read the Chapter Summary on page 426. You will find related material on the following pages: plant responses to moisture, 72-75; plant responses to light, 102-105; -3/2 power law, 400-402; genetic drift, 471-472; chemical defense in plants, 527-529; chemical defense in animals, 536; habitat fragmentation, 614-618.

Study Questions

1. What is meant by stress in populations? (412)
2. In what ways might animals, particularly mammals, respond physiologically to stress? (412)
3. How do some plants in crowded environment respond to competitive stress? (412-413)
4. What are pheromones? How can they function in population regulation among some mammals? (412)
5. Define dispersal. (413)
6. What types of animals are most prone to disperse? (413)
7. Distinguish between natal dispersal and breeding dispersal. (413)
8. When does most dispersal take place in the life cycle of an animal? (413)

9. In general, how far and in what manner do animals disperse? (413-415)
10. Why should animals disperse? (414)
11. What are some of the costs and benefits of dispersal? (415)
12. What is a population sink habitat? (416)
13. Can dispersal aid in population regulation and, if so, under what conditions? (416-417)

Key Terms and Phrases

dispersal

natal dispersal

breeding dispersal

presaturation dispersal

saturation dispersal

pheromones

stress

social dominance

territoriality

territory

floating reserve

home range

population sink

Key Term Review

1. Crowded conditions increase social contact within a population and lead to _____.

2. _____ _____ leads to a social organization or hierarchy based on intraspecific aggression.

3. Chemical substances released by an animal into the environment that can influence the behavior of others in the population are called _____.

4. Many animals avoid stress by seeking a less crowded area of their habitat. Such a movement away from a crowded area is called _____.

5. The area that an animal moves through regularly but does not defend is its _____ _____.

6. Young animals that leave their place of birth are involved in a _____ _____.

7. _____ refers to a behavior in which an individual claims an area for itself and excludes others from sharing it.

8. A defended area is a _____.

9. When adults leave a poor reproductive site for a better one they are involved in a _____ _____.

10. _____ _____ takes place before a population reaches its carrying capacity.

11. _____ _____ occurs when a population has exceeded its carrying capacity.

12. Individuals unable to claim a territory but capable of reproducing if a territory became available to them make up a _____ _____.

13. An unfilled habitat settled by dispersers that often exposes the animals to high predation or other causes of reproductive failure is a _____ _____.

Self Test

True and False

1. _____ As the size of a territory increases, the amount of energy expended in its defense increases.

2. _____ Stress can result in decreased births and decreased mortality of young.

3. _____ When a resource shortage causes dispersal to take place usually the reproductive adults are forced out.

4. _____ Social hierarchies become disruptive to normal social interactions.

5. _____ The size of a home range always remains the same.

6. _____ Increasing the number of aggressive encounters within a population will trigger changes in the hormone production of some individuals.

7. _____ Some individuals in a population are genetically predisposed to disperse.

8. _____ Territories are defended but home ranges are not.

9. _____ Males have smaller home ranges than females of the same species.

10. _____ The social structure of a population with alpha and omega individuals would be a social hierarchy.

11. _____ In mixed groups males and females may have their own social hierarchies.

12. _____ Peck orders in most social groups are relatively simple and linear.

13. _____ Dispersers are a random set of the resident population.

14. _____ As hypothesized, presaturation dispersal is a response to population density.

15. _____ If resources are unpredictable, it may be disadvantageous for an individual to be territorial.

Short Answer Questions.

1. Give two conditions under which benefits might exceed the costs of defending a territory.

2. Give one cost and one benefit for a young animal to remain on the home area. Then give one cost and one benefit for a young animal to leave a home area and settle elsewhere.

ANSWERS

Key Term Review

1. stress
2. social dominance
3. pheromones
4. dispersal
5. home range

6. natal dispersal
7. territoriality
8. territory
9. breeding dispersal

10. presaturation dispersal
11. saturation dispersal
12. floating reserve
13. population sink

Self Test

True and False

1. T
2. F
3. F
4. F
5. F

6. T
7. T
8. T
9. F
10. T

11. T
12. F
13. F
14. F
15. F

Short Answer Questions

1. Defending a territory can be costly both in energy and time. For benefits to exceed costs, the resource such as food or breeding habitat must be so distributed that it is easily defended. The acquisition of such resources must enable the territorial owner to have both the time and freedom from disturbance to carry on other activities such as feeding and caring for young.

2. One benefit among several that a young animal would gain by remaining in its home area would be familiarity with local terrain, reducing physical risks and improving chances for survival. However, it would face competition from kin for resources. One benefit accrued by moving would be avoidance of overcrowding and competition with kin, resulting perhaps in improved fecundity. A major cost would be increased vulnerability to predation, due in part to settling into unfamiliar terrain.

CHAPTER 20
LIFE HISTORY PATTERNS

Chapter Outline

Patterns of Reproduction
Mating Systems
   Monogamy
   Polygamy
Sexual Selection
   Models of Sexual Selection
   Processes of Selection

Reproductive Effort
   Parental Care
   Parental Investment
   Parental Energy Budgets
Gender Allocation
r and *K* Selection
Habitat Selection
Summary

Learning Objectives

After completing this chapter you should be able to:
- Distinguish between dioecious, monoecious, and hermaphroditic organisms.
- Describe various types of mating systems among animals.
- Discuss the models and process of sexual selection.
- Define reproductive effort and relate it to parental care and parental investment in offspring.
- Explain the relationship of age and size to fecundity.
- Discuss the relationship between parental investment in young and sex ratios in offspring.
- Describe the role of hermaphrodism to gender change in certain species of plants and fish.
- Distinguish between *r*-selection and *K*-selection.
- Discuss the role of habitat selection to reproductive success.

Summary

After reading this chapter and before continuing with the following material, read the Chapter Summary on pages 452-453. You will find related material on the following pages: adaptation, 30-31; energy allocation, 171-172; modular populations, 362-364; reproductive values, 384-387; population genetics, 456-477; natural selection, 461-463; fitness, 461; intraspecific predation, 541-542; parasitism, 558-577.

Study Questions

1. What forms can sexual reproduction take among plants and animals? (428)
2. What is a mating system? (428)
3. Define each of the following mating systems: polygamy, polygyny, polyandry, resource defense polygyny, female defense polygyny, male dominance polygyny, and resource defense polyandry. (429-430)
4. What is sexual selection and how might it function? (431-432)
5. What is the handicap hypothesis? (431)
6. Distinguish between intrasexual and intersexual selection. How are the two related? (432)
7. Distinguish between resource-based selection and genes-only selection. (432-435)
8. What is the difference between resource -based selection and genes-only selection? (433-435)
9 What is lek behavior? What three hypotheses have been advanced to explain lek behavior ? (435-436)
10. Define reproductive effort. (436)
11. Distinguish between precocial and altricial young and how do they relate to parental care? (436)
12. What are the energy costs of reproduction? How do individuals apportion energy between reproduction and other metabolic requirements? (437-440)
13. Distinguish between iteroparous and semelparous reproduction. (438)
14. What is the relationship between clutch size and litter size relative to energy apportioned in reproduction? (441-443)
15. What is hermaphrodism? Sequential hermaphrodism? How is it involved in gender change? (443-445)
16. How does gender change maximize individual reproductive success? (444-445)
17. What is the relationship between age, size, and fecundity? (445-446)
18. How might resource allocation influence sex ratios in a population? (446-448)
19. Distinguish between r-selection and K-selection. (448-449)
20. What are r-traits and K-traits? What is the weakness of the concept? (448-449)
21 How might animals select an optimal habitat? (450-452)

Key Term and Phrases

dioecious
monoecious
hermaphroditic
mating system
monogamy
polygamy
polygyny
polyandry
promiscuity
resource-defense polygyny
female-defense polygyny
male-dominance polygyny
resource-defense polyandry
sexual selection
handicap hypothesis

intersexual selection
resource-based selection
genes-only selection
lek
female choice
hotspot hypothesis
hotshot hypothesisl
precocial
altricial
semelparous
iteroparous
optimal lifetime reproductive success
r-strategist
K-strategist
reproductive effort

intrasexual selection   habitat selection
explosive breeding assembly

Key term Review

1. The amount of time and energy an organism puts into reproduction is called
    _____ _____.

2. After hatching, fledgling robins are helpless, featherless, and must be fed by their
    parents. They are _____ young.

3. _____ are generally associated with characteristics as small body
    size, high rate of reproduction, and rapid development.

4. Plants possessing bisexual flowers with both male (stamens) and female (ovules)
    organs are called _____.

5. Grazing animals, whose young can move about and forage for themselves shortly
    after birth, produce _____ young.

6. _____ have both female and male sex organs in a individual.

7. _____ generally are characterized by a large body size.
    slow rate of reproduction, and slow development.

8. _____ involves a pair bond between one male and one female.

9. Salmon swimming upstream from the ocean to engage in one massive reproductive
    effort and then dying is an example of a _____ organism.

10. Organisms with repeated reproduction throughout a lifetime are _____,

11. Bull elk battling other elk for the opportunity to mate with females is _____
    _____.

12. Females making their own choice of a mate is _____ _____.

13. In lek behavior females showing a preference for a courtship arena are exhibiting
    _____ _____.

14. Bull elk controlling or gaining access to a group of females engage in
    _____ _____ _____.

15. Males gaining control of or access to two or more females is _____.

16. Females gaining control of or access to two or more males is _____.

17. A _____ _____ involves the manner in which males and females
    acquire mates, the nature of the pair bond, and the pattern of parental care.

18. Males and females that mate without establishing a pair bond are engaging in
_____.

19. Prairie chickens congregating on a piece of ground to attract females is an example
of _____ _____ _____.

20. The display ground is called a _____.

21. Females that defend resources essential to males and require the males to brood
the young are practicing _____ _____ _____.

22. Individuals of both sexes that congregate for a short-lived, highly synchronized
mating period are engaged in an _____ _____
_____.

23. When hummingbirds obtain mates by controlling access to a nectar food supply,
they are engaged in _____ _____ _____.

24. In _____ _____ females mate only with the highest
ranking males.

25. Females that choose a male based on the quality of his territory or the defended
food supply are employing _____ _____.

26. According to the _____ _____, males congregate on leks in areas
where encounters with females are exceptionally high.

27. The _____ _____ proposes that because of a strong social hierarchy
among males, dominant males leave little opportunity for female choice of a mate.

28. Individuals making a choice of a mate are engaged in _____ _____.

29. The _____ _____ holds that females prefer males with very bright
plumage or other secondary sexual characteristics that could reduce his survival;
his survival is proof of a superior genotype.

30. The acquisition of two or more mates, none of which is mated to other individuals, is
_____.

31. The sum of present and future reproductive success is the _____
_____ _____ _____.

32. The process by which animals cue in on a place to settle is _____
_____.

Self Test

True and False

1. _____ Among homeotherms fecundity increases with size and age.

2. _____ The evolution of dioecy from male-dominated hermaphroditic flowers is evidence of sexual selection among plants.

3. _____ Oak trees produce a few large seeds containing large amounts of stored energy. They reflect the characteristics of plants adapted to harsh environments.

4. _____ Reproductive costs involve a tradeoff between the number of present offspring and future offspring.

5. _____ Hermaphrodites maintain genetic variability by self-fertilization.

6. _____ Parental care is well developed in salamanders.

7. _____ Internal fertilization and terrestrial reproduction is well developed in reptiles.

8. _____ Most vertebrates are iteroparous.

9. _____ Among iteroparous organisms, early reproduction means a reduced potential for future reproduction.

10. _____ Among poikilotherms, larger fecundity means poorer parental survival.

11. _____ Plants growing in adverse environments allocate more energy to reproduction that those living in more moderate environments.

12. _____ Among poikilotherms fecundity increases with size and age.

13. _____ The sex ratio in a population is skewed to the less expensive sex.

14. _____ Among some species of plants, individuals may be a male one year and a female the next.

15. _____ Females have a greater variability in reproductive success than males.

16. _____ The absence of an individual of one sex in certain fish populations can stimulate gender change in some individuals of the other sex.

17. _____ Evidence exists that suggests that among birds, males with more complex song patterns are most successful in attracting females.

18. _____ Among some fish females prefer to mate with males already defending a nest.

19. _____ A method of brood reduction comes about by siblicide.

20. _____ Clutch and litter size among birds and mammals are greater in temperate regions than in tropical regions.

21. _____ Most hermaphrodites are self-fertilized.

22. _____ Plants of higher latitudes allocate more resources to reproduction than plants of lower latitudes.

23. _____ Vegetation structure is one key to habitat selection.

24. _____ Although some habitats are physically suitable, organisms may avoid them because other individuals of their own species are not present.

Matching

According to theoretical ecology, species may be *r*-strategists or *K*-strategists. Each has certain characteristics. Identify each of the following characteristcs by placing *r* or *K* on the line next to the statement.

1. _____ Ability to disperse over a wide area.

2. _____ Populations often widely fluctuating.

3. _____ Density-independent mortality.

4. _____ Slow maturation.

5. _____ Unable to colonize highly disturbed sites.

6. _____ Production of many small seeds.

7. _____ Invest considerable energy in parental care.

8. _____ Populations are often at or near carrying capacity.

Short Answer Questions

1. What are three hypotheses advanced to explain latitudinal variation in clutch size?

2. What are advantages of self-fertilization in plants?

166

Read the following narrative carefully. Questions testing your ability to apply concepts to the situations described will appear at the end.

In late March the first red-winged blackbirds returned to the marsh. One of the early arriving males settled onto a prime area of the marsh where stands of cattails would grow dense and the potential for insect abundance was high. The male, whom we will call A, proclaimed his ownership of that part of the marsh by song and display of his red shoulder patches

Within a week more males arrived on the marsh. Male A, unable to defend the large area he staked out, had to pull in his territorial boundaries, but he still retained his area of prime habitat. About two weeks later females returned to the marsh. One female, whom we will call C, attracted to male A, settled on his territory, mated and began building a nest in the cattails. Later a second female, D, also attracted to A's prime territory, failed to settle with a male occupying a territory on the edge of the marsh. She moved into A's territory, mated with him, built a nest, and reared a brood.

Both C and D had clutches of 4 eggs. Upon hatching the young lacked feathers and needed parental feeding. Although male A helped C more than D in feeding the young, both C and D successfully reared broods, largely because of the abundant food supply in A's territory and because C's brood hatched earlier than D's.

In late summer the parents and young assembled in large flocks and fed beyond the marsh. Territorial boundaries were forgotten, but the birds established social interactions among them involving social dominance. By late fall, the birds migrated south. A, C, and D would return to the marsh next spring to nest again.

In the same marsh lived two pairs of muskrats, each of whom defended their own area of the marsh, built lodges of cattail stems, and reared young. Pair 1 produced 16 young in 3 litters. Pair 2 produced 10 young in 2 litters. The naked young grew rapidly; they were furred and swimming at two weeks of age. Within four weeks the parents drove the young from their area prior to producing the next litter. The young settled in unoccupied areas of the marsh until fall. At that time many young left the marsh and settled along streams and in farm ponds. Several young remained behind and established territories in the home marsh. The owner of one farm, fearing the muskrats might damage the dam, trapped the muskrats as they moved into the pond. No muskrats were able to become established there.

In brushy fields adjacent to the marsh live bobwhite quail. The males in spring claim areas of hayfields and brushy lands which each defends from other male bobwhites and to which each attempts to attract a mate by whistling "bob-white."

The female, attracted to the male's territory, mates and about 50 meters from the edge of brushy field, builds a nest of dried grasses in an open field. She lays and incubates a clutch of 15 eggs that hatch in 23 days. Upon hatching the downy young run about as quickly as they dry off. The female and male call softly to the chicks who follow the parents through grass and brush and feed on an abundance of insects.

Not all male bobwhites, however, are successful at attracting a mate and spend the summer in brushy areas. Occasionally they attempt to make advances toward the mated females, but are driven away by the male. On several territories, a mated male disappeared. In each case, his place was taken by one of the unmated birds.

Select the concepts listed below as matching items that best fit the statements based on the narrative. You will not use all terms given and you may use some more than once.

A. Semelparity
B. Precocial
C. Floating reserve
D. Contest
E. Female defense polygyny
F. Altricial
G. Territoriality
H. Dispersal
I. Resource-based selection

J. Iteroparity
K. Home Range
L. Scramble
M. Territory
N. Population sink
O. Monogamy
P. Social hierarchy
Q. Resource defense polygamy
R. Genes-only selection

1. _____ Possible population regulatory mechanisms exhibited by the behavior of male red-winged blackbirds.

2. _____ Type of young produced by red-winged blackbirds.

3. _____ Type of reproductive strategy employed by red-winged blackbirds, muskrats, and bobwhite quail.

4. _____ Area defended by red-winged blackbird.

5. _____ Type of intraspecific competition involved in the social relations of the muskrats on the marsh.

6. _____ Type of mating systems found in the red-winged blackbird.

7. _____ Type of young produced by the muskrats.

8. _____ Social relationships exhibited by red-winged blackbirds in the fall.

9. _____ Behavior exhibited by young muskrats.

10. _____ What did the farm pond become for the young muskrats?

11. _____ Type of young produced by bobwhite.

12. _____ Role of unmated birds in the bobwhite population.

13. _____ Type of mating system found in the muskrats.

14. _____ Type of mating system employed by the bobwhite quail.

ANSWERS

Key Term Review

1. reproductive effort
2. altricial
3. r-strategists
4. monoecious
5. precocial
6. hermaphrodites
7. K-strategists
8. monogamy
9. semelparity
10. iteroparous
11. intrasexual selection
12. intersexua l selection
13. female choice
14. female defense polygyny
15. polygyny
16. polyandry
17. mating system
18. promiscuity
19. male dominance polygyny
20. lek
21. resource defense polyandry
22. explosive breeding assembly
23. resource defense polygyny
24. genes-only selection
25. resource-based selection
26. hotspot hypothesis
27. hotshot hypothesis
28. sexual selection
29. handicap hypothesis
30. polygamy
31. optimal lifetime reproductive success
32. habitat selection

Self Test

True and False

1. F	9. T	17. T
2. T	10. F	18. T
3. F	11. T	19. T
4. T	12. T	20. F
5. F	13. T	21. F
6. F	14. T	22. T
7. T	15. F	23. T
8. T	16. T	24. T

Matching

1. r
2. r
3. r
4. K
5. K
6. r
7. K
8. K

Short Answer Questions

1. One, the daylength hypothesis, holds that longer daylength at higher latitudes enables parents to forage longer for food to support large broods. In tropics where daylength does not change, there is less time to forage. A second hypothesis relates to competition and predation. In the low latitudes birds face more competition for food all year long because of the presence of winter migrants. In higher latitudes birds have less competition and more abundant food. Likewise there appears to be more predation pressure in the tropics. A third hypothesis relates to seasonal variation in food supply used by a population. The breeding population is regulated by winter mortality. Greater winter mortality means more food available during the breeding season. This resource availability is reflected in larger clutch sizes.

2. Self-fertilization ensures some seed production should cross-fertilization fail during any one season. It also enables a single, self-fertilized plant to colonize a new habitat, reproduce itself, and establish a new population. Self fertilization also ensures a population that is uniquely adapted to local environmental conditions.

Comprehensive Test

1. G	8. O
2. F	9. H
3. J	10. M
4. L	11. B
5. D	12. C
6. P	13. N
7. F	14. I

CHAPTER 21
POPULATION GENETICS

Chapter Outline

Genetic Variation
  Types of Variation
  Genotypes and Phenotypes
  Sources of Variation
  Hardy Weinberg Equilibrium
Natural Selection
  Nonrandom Reproduction
  Fitness and Modes of Selection
  Group and Kin Selection

Inbreeding
  The Inbreeding Coefficient
  Consequences of Inbreeding
Genetic Drift
  Effective Population Size
  Gene Flow, Mating Strategies
    and Genetic Driftt
Minimum Viable Populations
Summary

Learning Objectives

After completing this chapter you should be able to:
- Describe the types of variations in a population.
- Distinguish between a genotype and a phenotype.
- Discuss the sources of genetic variation.
- Explain the Hardy-Weinberg Equilibrium and its importance to population genetics.
- Discuss natural selection, fitness, and types of selection.
- Distinguish between group and kin selection.
- Define inbreeding and discuss its consequences.
- Explain genetic drift and distinguish it from inbreeding.
- Discuss the concept of effective population size.
- Discuss the meaning of minimum viable populations and its application to small populations.

Summary

After reading this chapter and before continuing with the material below, read the Chapter Summary on pages 476-477. You will find related material on the following pages: hypothesis testing, 18-21; adaptation, 30; age structure, 372-374; population growth 392-396; extinction, 407-409; demographic stochasticy, 408; environmental stochasticity, 408; habitat fragmentation, 614-618.

Study Questions

1. What are the types of variation among individuals of a population? (456)
2. Distinguish between genotypes and phenotypes. (456-457)
3. What are the major sources of variation? (457-458)
4. What is phenotypic plasticity? (457)

5. What is the meaning of the following: diploid, haploid, mitosis, meiosis, genes, alleles, homozygous, heterozygous, and polyploidy? (457)
6. What are the major sources of genetic variation? (457-458)
7. What is a mutation? Distinguish between a macromutation and a micromutation. (458)
8. What is the Hardy-Weinberg Equilibrium? (458-459)
9. What is the formula for gene frequency? Genotypic frequency? (460)
10. What conditions must be met for Hardy-Weinberg equilibrium? (460)
11. What is natural selection? (461)
12. What are the three main kinds of selection? What is a selection coefficient? (462)
13. What is group selection? Kin selection? (463-467)
14. What is altruism? Inclusive fitness? How are the two related? (46-67)
15. What is inbreeding? (467)
16. Why does inbreeding increase homozygosity? (468)
17. Define the coefficient of inbreeding. What does it mean? (468-469)
18. Contrast allozygous and autozygous genes? (468)
19. What are the end values of F, the inbreeding coefficient and what do the values tell us? (469)
20. What is inbreeding depression? Outbreeding depression? (470)
21. To what degree and under what conditions does inbreeding occur in natural populations? (470)
22. What is genetic drift? (471)
23. If inbreeding is involved in genetic drift, what is the difference between inbreeding and genetic drift relative to mating? (471)
24. What is meant by effective population size? How does it relate to inbreeding? (472)
25. What is the formula for calculating effective population size? (472-473)
26. How do sex ratios of reproducing individuals in polygamous populations influence effective population size? (472-473)
27. Show that a breeding populations with a ratio of 4 males to 10 females is less than a population of 6 males and 10 females. (472)
29. What is a population bottleneck? What is its significance genetically? (474)
30. What is the founder effect? (474)
31. What relationship exists among dispersal, mating patterns, and genetic drift? (474-475)
32. What is meant by a minimum viable population? (475-476)

Key Terms and Phrases

gene pool	mutation	group selection
discontinuous variation	macromutation	deme
continuous variation	micromutation	inclusive fitness
genotype	polyploidy	kin selection
phenotype	deletion	eusociality
phenotypic plasticity	duplication	inbreeding
diploid	inversion	inbreeding coefficient
meiosis	translocation	allozygous
mitosis	genotypic frequency	autozygous
haploid	allele frequency	inbreeding depression
diploid	fitness	genetic drift
genes	selection coefficient	outbreeding depression

homozygous   stabilizing selection   fixation index
heterozygous   directional selection   effective population size
recombination   disruptive selection   bottleneck
allele   minimum viable population   founder effect

Key term Review

1. Alleles identical by descent are _____.

2. A _____ occurs when a population experiences catastrophic decline and eventual recovery.

3. A small local population is a _____

4. One of two possible forms of a gene at a given locus is an _____.

5. A _____ _____ is the total genetic variability in a population.

6. A mixing of alleles in a population though sexual reproduction is _____.

7. The external expression of a genotype is the _____.

8. The _____ _____ is a measure of increase in homozygosity in inbred populations.

9. The type of selection that favors the average is _____ _____.

10. The sum of hereditary information carried by an individual is the _____.

11. Sharp differences in color in a population is an example of a _____ _____; a range of heights in a population is an example of a _____ _____.

12. The ability of a genotype to give rise to different phenotypic expressions under different environmental conditions is _____ _____.

13. During cell reproduction, called _____, the cell retains a full complement of chromosomes, called the _____ number.

14. In the production of gametes, the process of cell division is called _____; the pairs of chromosomes are split, resulting in one-half of a full complement called the _____ number.

15. Units of heredity carried by the chromosomes are _____.

16. If each member of the pair of alleles affect a given trait in the same manner, the pair is called _____. If each member affects a trait in a different manner, the pair is called _____.

17. An inheritable change in a gene or a chromosome is a _____.

18. A duplication of entire sets of chromosomes is _____.

19. The proportion of alleles in a population is the _____ _____; the proportion of pairs of alleles in a population is the _____ _____.

20. Chromosomal mutations are _____. These include _____, a loss of part of a chromosome; _____, the internal doubling of a chromosome; _____, the alteration of the sequence of genes in a chromosome; and _____, the exchange of segments between two nonpaired chromosomes.

21. Gene mutations are _____.

22. The measure of an individual's contribution to future generations is _____.

23. The selective pressure acting on a genotype is designated as a _____ _____.

24. Selection that favors one extreme phenotype over others is _____, whereas selection that favors both extremes over a range of phenotypes is _____.

25. _____ _____ operates on the differential productivity of local populations.

26. Differential reproduction among groups of closely related individuals is _____ _____.

27. Homozygous alleles occupying a given locus that are identical because of independent mutations and therefore are not identical by descent are _____.

28. The sum of the total fitness of an individual and the fitness of its relatives, weighted toward the degrees of relationship is _____ _____.

29. Mating among close relatives is _____.

30. The detrimental effects of inbreeding are called _____ _____.

31. A loss of fitness that comes about through the contamination of a gene pool with new alleles that produce offspring poorly adapted to the local environment is _____ _____.

32. A form of altruism, widespread in insect populations, characterized by extreme social involvement by members of a population including cooperative care of young and division of labor is _____.

33. A measure of heterozygosity that has been lost through inbreeding in a population is the _____ _____.

34. A measure of genetic drift is the _____ _____.

35. The chance fluctuation of allele frequencies in small populations as a result of random sampling is _____ _____.

36. A small population carrying only a sample of the genes from a parent population introduced in a new or empty habitat is subject to random genetic drift known as _____ _____.

37. The size of an ideal population that would undergo the same amount of random genetic drift as the actual population is the _____ _____ _____.

38. The threshold number of individuals that will ensure the persistence of a subpopulation in a viable state for a given number of years is the _____ _____ _____.

Self Test

True and False

1. _____ Variations in characteristics that result from disease or injury are inheritable.

2. _____ Individual fitness is measured by the number of reproducing offspring it leaves.

3. _____ Genetically-based advantageous phenotypes tend to increase in frequency.

4. _____ Individuals in a population with more adaptive traits have a greater chance of leaving more offspring.

5. _____ Polyploidy is the duplication of sets of genes.

6. _____ A major source of genetic variability is mutation.

7. _____ Traits promoted by individual selection are considered selfish.

8. _____ Fitness acquired through parental reproduction is inclusive fitness.

9. _____ An altruistic act means a sacrifice of life.

10. _____ Group selection, if it occurs, would have to operate on differential productivity of local populations.

11. _____ Kin selection involves closely related individuals.

12. _____ Cost/benefit of an altruistic act is greater if the recipients are cousins rather than full sibs.

13. _____ The concept of kin selection seems to explain many aspects of social behavior.

14. _____ The diploid state, *Aa,* is heterozygous.

15. _____ Evolution is a change in gene frequency through time.

16. _____ Genetically based advantageous phenotypes tend to increase in frequency.

17. _____ Phenotypic plasticity refers to variations in phenotypic expressions as a response to varying environmental conditions.

18. _____ When and where a mutation occurs depends upon chance.

19. _____ Natural selection acts on an organism's dominant alleles only.

20. _____ Selection will not eliminate a lethal recessive allele from a large population of diploid organisms because there will always be some heterozygous carriers for the allele.

21. _____ Genetic drift increases heterozygosity.

22. _____ Alleles that are identical by descent are allozygous.

23. _____ As the ratio between males and females widens in a polygamous population, the effective population size declines.

24. _____ Bottlenecks occur when a large population is drastically reduced in size.

25. _____ In monogamous species, the breeding population equals the effective population size.

26. _____ The inbreeding coefficient defines the loss of heterozygosity.

Which of the following assumptions meet (**T**) or do not meet (**F**) the Hardy-Weinberg requirements for genetic equilibrium in a population:

27. _____ Random mating cannot occur.

28. _____ No mutations may occur or can be accounted for.

29. _____ No migration may occur.

30. _____ Natural selection may occur.

31. _____ Population is small.

32. _____ Generations overlap.

33. _____ Reproduction is asexual.

Relative to the effective population size, which of the following assumptions apply to the ideal population? Place **T** for applicable assumption, **F** for the nonapplicable.

34. _____ Population growth of the ideal population is the same as that of the population under study.

35. _____ Nonrandom breeding.

36. _____ A 1 : 1 sex ratio.

37. _____ The number of progeny per family is randomly distributed.

Matching

Match these terms with the statements below. Some terms will not be used,.

A. Genetic drift   F. Random   K. Homozygous
B. Inbreeding   G. Founder principle   L. Bottleneck
C. *F.*   H. Autozygous   M. Heterozygous
D. Allozygous   I. Polygamous   N. N_e
E. Nonrandom   J. Monogamous

1. _____ Alleles identical by descent.

2. _____ Population experiences catastrophic decline and eventual recovery.

3. _____ Parent-sib mating.

4. _____ Isolated small population experiences this.

5. _____ Measure of increase in homozygosity in inbred populations.

6. _____ A small group of wild turkey transplanted into a new habitat carry only a fraction of the total genetic variation of the parental population.

7. _____ A measure of effective population size.

8. _____ Inbreeding is alleviated more rapidly by immigration in a population with this type of breeding system.

9. _____ Retains rare alleles in a population.

10. _____ Genetic drift experiences this type of inbreeding.

Associate each of the statements with their appropriate terms:

A. Genotype   E. Genotypic frequency   I. Stabilizing
B. Fitness   F. $\sqrt{q^2}$ J. Recombination
C. Gene pool   G. Heterozygosity   K. Deme
D. Homozygous   H. Gene frequency   L. Directional
                                                     M. Phenotype

11. _____ Total genetic variability in a population.

12. _____ Process by which alleles in a population are mixed.

13. _____ What natural selection works on.

14. _____ q.

15. _____ Type of selection that favors the average.

16. _____ Possession of different alleles at same locus.

17. _____ $p^2 + 2pq + q^2 = 1$.

18. _____ A small local population.

19. _____ $p + q = 1$.

20. _____ Comparative ability of individuals to leave behind reproducing offspring in the next generation.

Problem Questions

1. In a population sample of 500 individuals, 125 were homozygous for *yy*.

   A. What is the genotypic frequency of yy? .

   B. What is the allele frequency of y?

2. Given a genotypic frequency of 0.36 for *bb*

   A. What is the gene frequency of *b*?

   B. What is the gene frequency of *BB*

3. Given a value of $p = 0.6$, what is the genotypic frequency of alleles *AA*, *Aa*, and *aa*?:

4. If F, the fixation index or measure of decline of heterozygosity, of population A is 0.06 and of population B is 0.25, which population is experiencing the most genetic drift?.

5. If the inbreeding coefficient, F, for individual A is 0.25 and for individual B is 0.50, which individual is the most highly inbred?

6. Two populations, A, which is monogamous, and B, which is polygamous, both have 100 individuals in the breeding population. Population A has 50 males and 50 females. Population B has 20 males and 80 females. Which of the following is the effective population of A and the effective population size of B: **100, 50, 75, 64, 80, 20**?

Short Answer Questions

1. What is a population bottleneck and what are the genetic effects on the affected population?

2. When inbred individuals disperse from the population and join another they may experience outbreeding depression? What is outbreeding depression and how does it affect a population?

3. What are three outcomes of inbreeding (called inbreeding depression)?

4. Both inbreeding and genetic drift involve an increase in the frequency of homozygous alleles. What is the major difference between the two? In other words, how does inbreeding in a small population differ from genetic drift in a small population?

ANSWERS

Key Term Review

1. autozygous
2. bottleneck
3. deme
4. allele
5. gene pool
6. recombination
7. phenotype
8. inbreeding coefficient
9. stabilizing selection
10. genotype
11. discontinuous variation
    continuous variation
12. phenotypic plasticity
13. mitosis, diploid
14. meiosis, haploid

15. genes
16. homozygous
    heterozygous
17. mutation
18. polyploidy
19. allele frequency
    genotypic frequency
20. macromutations
    deletion
    duplication
    inversion
    translocation
21. micromutation
22. fitness
23. selection coefficient

24. directional
    disruptive
25. group selection
26. kin selection
27. allozygous
28. inclusive fitness
29. inbreeding
30. inbreeding depression
31. outbreeding depression
32. eusociality
33. inbreeding coefficient
34. fixation index
35. genetic drift
36. founder effect
37. effective population size
38. Minimum viable
    population

Self Test

True and False

1. F	11. T	21. F	31. F
2. T	12. F	22. F	32. F
3. T	13. T	23. T	33. F
4. T	14. T	24. T	34. T
5. T	15. T	25. T	35. F
6. F	16. T	26. T	36. T
7. T	17. T	27. F	37. T
8. F	18. T	28. T	
9. F	19. F	29. T	
10. T	20. T	30. F	

Matching

1. H	11. C
2. L	12. J.
3. B	13. M
4. A	14. F
5. C	15. I
6. G	16. G
7. N	17. E
8. I	18. K
9. M	19. H
10. F	20. B

Problem Questions

1. A. The genotypic frequency of *yy* is 0.25. The genotyptic frequency is expressed as $p^2 + 2pq + q^2 = 1$. In the population of 500, *yy* was represented in 125 individuals. Its proportion in the population is 125/500 = 0.25, the value of q^2

   B. The frequency of the allele *y* is obtained by $\sqrt{q^2}$; $\sqrt{0.25}$ = 0.50; q = 0.50. The frequency of allele y is 0.50.

2. A. The genotypic frequency of *bb* is 0.36; thus the gene frequency of *b* is $\sqrt{0,36}$ = 0.60.

   B. Because $p + q = 1$, the gene frequency of *B* is 1 - q = 1 - 0.60 = 0.40.

3. The genotypic frequency is expressed by $p^2 + 2pq + q^2 = 1$. The gene frequency for A (p) is 0.6. Therefore the genotypic frequency for AA (p^2) is $(0.6)^2 = 0.36$. The genotypic frequency for aa is $(1 - 0.6)^2$ or $(0.4)^2 = 0.16$. The genotypic frequency for Aa ($2pq$) is $2(0.6 \times 0.4) = 0.48$.

4. The value of F, the fixation index, ranges from 0, no homozygosity or complete heterozygosity to 1, complete homozygosity or no heterozygosity. Therefore B with a fixation index of 0.25 is experiencing the most genetic drift.

5. The inbreeding coefficient, F, ranges from 0, indicating no inbreeding, to 1, complete inbreeding. Because the inbreeding of B is 0.50 is higher than that of A, B exhibits more inbreeding.

6. The effective population size of monogamous population A is 100. The sex ratio is 1 to 1. The effective population size of the polygamous population B is 64. This value is obtained by the equation for effective population size;

$$N_e = 4N_m N_f / N_m \, N_f \, ; \quad 4(20 \times 80)/20 + 80 \; = \; 6400/100 = 64$$

Short Answer Questions

1. In a population bottleneck, the population experiences a drastic drop in numbers. The remaining population carries only a sample of the gene pool of the prebottleneck population. Some genes may be lost from the population, reducing genetic diversity. As the remnant population grows, it also experiences genetic drift and increased homozygosity.

2. New individuals introduced into a population bring new genes into the gene pool. These genes may be maladapted to the new environment. Offspring may be poorly adapted to the local environment, reducing overall fitness. This loss of fitness is outbreeding depression.

3. Three outcomes of inbreeding (called inbreeding depression) are: decreased fertility, small body size, loss of vigor. (For others see text, p. 470).

4. Both inbreeding and genetic drift experience an increase in homozygous alleles. The difference between the two is: in genetic drift the increase in homozygosity comes from random mating, whereas in inbreeding, mating is nonrandom between very closely related individuals.

CHAPTER 22
INTERSPECIFIC COMPETITION

Learning Objectives

After completing this chapter you should be able to:

- Briefly describe the various types of interspecific relationships.
- Define interspecific competition.
- Discuss the possible outcomes of interspecific competition.
- Discuss the competitive exclusion principle.
- Explain what is meant by diffuse competition.
- Discuss the approaches to the study of competition.
- Discuss the evidences for coexistence and exclusion.
- Explain the concept of resource partitioning.
- Discuss the concept of the niche, including the fundamental and realized niche.
- Distinguish between niche width, niche overlap, niche release, and niche shift.

Summary

After reading this chapter and before continuing with the material below, read the Chapter Summary on page 499. You will find related material on the following pages: trophic levels, 191-192; food chain, 185-187, 192-193, 336, 621-623; population growth, 392-396; intraspecific competition, 412-424; social dominance, 417-418; territoriality, 418-425; chemical defense by plants, 527-529; parasitism, 558-573; competition and community structure, 620-621; predation and community structure, 621-623; plant succession, 656-669.

Study Questions

1. Distinguish between the following population interactions: neutral, mutualism, commensalism, amensalism, parasitism, competition, and predation. (480)
2. According to the Lotka-Volterra theory of competition, what are the four possible outcomes of competition? (480-482)
3. What is competitive exclusion? What conditions are necessary for competitive exclusion? (482)
4. What is meant by diffuse competition? (482, 489)
5. Why is it difficult to demonstrate interspecific competition in the field and much easier in the laboratory? Can you apply lab study results to field situations? (484-488)
6. Why is coexistence among competing species possible? (486-487)
7. What is allelopathy? (489)
8. What is meant by resource partitioning? (489-491)
9. What is the importance of differential resource use to interspecific competition?
10. Define the niche. What is the functional niche? The realized niche? (494-495)
11. What is niche overlap? Niche width? Niche change? (495-499)

Key Terms and Phrases

mutualism
commensalism
amensalism
predation
parasitism
interspecific competition
exploitative competition
interference competition
Gause principle
competitive exclusion principle
realized niche

diffuse competition
niche
hypervolume
fundamental niche
niche overlap
niche width
niche shift
ecological release
niche compression
allelopathy

Key Term Review

1. When two or more species use a portion of the same resource simultaneously, they are experiencing _____ _____.

2. The set of environmental conditions a population actually uses is its _____ _____.

3. _____ _____ occurs when two or more species use the same resource in short supply.

4. The production and release of chemicals by plants that inhibit the growth of other species is called _____.

5. The _____ defines the functional role of a species and the habitat it occupies.

184

6. _____ is an interaction between two species in which one benefits and the other receives no benefit or harm.

7. A _____ _____ is the total range of conditions under which a species can survive.

8. This multidimensional niche is called the _____.

9. _____ involves a one-sided relationship between two species in which one species is harmed and the other is neither benefited or harmed.

10. _____ is the killing and consumption of prey.

11. The _____ _____ _____ states that that two species that are complete competitors cannot coexist. This concept is also called the _____ _____.

12. In _____ _____ a competitor is denied access to a resource.

13. An interaction in which a small organisms lives in and derives its nourishment from another usually larger organism is _____.

14. In _____ _____ the use of a resource by one species reduces its availability for another.

15. The sum of weak competitive interactions among ecologically related organisms is _____ _____.

16. Competition that results in a contraction of an organism's habitat rather than the type of resources used is _____ _____.

17. Niche expansion in response to reduced interspecific competition is called _____ _____.

18. Adoption of changed feeding or other behavioral patterns by two or more competing populations to reduce interspecific competition is called a _____ _____.

19. An interspecific relationship in which both species benefit is called _____.

20. The range of a single niche dimension occupied by a population is ts _____ _____.

Self Test

True and False

1. _____ It is unlikely that two species will have exactly the same niche requirements.

2. _____ In the Lotka-Volterra competition equations, competition coefficients represent a drag on the growth of one population by another.

3. _____ One outcome of interspecific competition is an unstable equilibrium between populations.

4. _____ Most species experience direct competition rather than diffuse competition.

5. _____ The position occupied by a species on a resource gradient is one aspect of its niche.

6. _____ Resource partitioning is a likely outcome of intraspecific competition.

7. _____ Organisms may undergo niche shift and thereby reduce competition.

8. _____ An individual free from competition or interference from other organisms would occupy its realized niche.

9. _____ Two complete competitors cannot coexist; one will go extinct.

10. _____ Species with broad niches are specialists.

11. _____ If competing species share a resource, coexistence results, even if the competition reduces the fitness of both populations.

12. _____ Each species usually has a greater requirement for one type of resource than for another.

13. _____ Interspecific competition is probably continuous.

14. _____ Important in interspecific competition is the rate of consumption vs. the rate of renewal of a resource.

15. _____ Commensalism is an interaction between two species in which neither benefits.

16. _____ Competitive exclusion is most conspicuous among native plants.

17. _____ Diffuse competition is an example of indirect competition.

18. _____ Intraspecific competition can be more influential in a population than interspecific competition.

Match the statements below with the following terms: (Not all of the terms will be used.)

A. Ecological release   D. Interference competition   G. Niche shift
B. Exploitation competition   E. Resource partitioning   H. Niche width
C. Niche compression   F. Diffuse competition   I. Realized niche

1. _____ Occupying the same habitat, species A feeds on larger-sized seeds and Species B feeds on smaller-sized seeds.

2. _____ Species A and Species B both share the same resources which can be in short supply.

3. _____ Occupying the same habitat, Species A and Species B overlap on the larger gradient of seed sizes eaten; Species B and Species C also overlap on the lower gradient of seed sizes. The competition between A and B reduces the population of B, allowing the population of C to expand.

4. _____ Species B expands its habitat when the local population of Species A declines.

5. _____ Species A changes its feeding pattern in response to competition from Species B.

6. Interspecific relationships can be considered positive, negative, or neutral as they affect individualsof both species involved. Using **+** for positive interactions, **-** for negative interactions, and **0** for neutral effects, fill in the following table indicating the effects of the relationships on individuals of the two species involved.

Relationship	Species A	Species B
Mutualism		
Commensalism		
Predation		
Parasitism		
Competition		

Short Answer Questions

1. Why is competitive exclusion most apparent between native and exotic species in the same habitat?.

2. Why is the unstable or stable coexistence rather than exclusion the norm among species?

3. What are some of the problems encountered trying to demonstrate interspecific competition in the field?

Answers

Key Term Review

1. niche overlap
2. realized niche
3. interspecific competition
4. allelopathy
5. niche
6. commensalism
7. fundamental niche
8. hypervolume
9. amensalism
10. predation
11. competitive exclusion principle
    Gause's principle
12. interference competition
13. parasitism
14. exploitative competition
15. diffuse competition
16. niche compression
17. ecological release
18. niche shift
19. mutualism
20. niche width

Self Test

True and False

1. T	7. T	13. F
2. T	8. F	14. T
3. T	9. T	15. F
4. F	10. F	16. F
5. T	11. T	17. T
6. F	12. T	18. T

Matching

1. E
2. B
3. F
4. A
5. G

Relationship	Species A	Species B
Mutualism	+	+
Commensalism	+	0
Predation	+	-
Parasitism	-	+
Competition	-	-

Short Answer Questions

1. Native species of plants in a given environment have established a competitive relationship over a long period of time. If an exotic species well adapted to local environmental conditions invades the area, it has in effect escaped its old competitors and the native species cannot resist this new competitor. Such invaders (such as loosestrife, Japanese honeysuckle, and multiflora rose) are aggressive, have a high growth rate, high seed production, and outcompete native species for space, light, and nutrients.

2. Stable and unstable coexistence is the norm rather than the exception for several reasons. Environmental conditions are variable. At one set of conditions one species may have the competitive advantage. As conditions change the competitive advantage may switch to the other species. Resource levels may vary providing abundance or scarcity for one species relative to another. Intraspecific competition may be more important to the competing species than interspecific competition.

3. Field experiments to demonstrate interspecific competition have many problems. Short term studies cannot demonstrate whether competition does or does not exist over a long time period. Variable environmental conditions and changes in levels of population abundance affect interspecific relationships. Replication, design, and timing impose problems that may be difficult to overcome.

CHAPTER 23
PREDATION

Learning Objectives

After completing this chapter you should be able to:
- Discuss the strengths and weaknesses of the Lotka-Volterra, Nicholson-Bailey and MacArthur-Rosenzweig models of predation.
- Describe some laboratory and field studies of predator and prey systems.
- Explain functional response and distinguish between Type I, Type II, and Type III functional responses.
- Discuss the concepts of threshold of security, aggregrative response, switching, and search image as they relate to functional response.
- Describe numerical response.
- Discuss the various aspects of the foraging theory.
- Distinguish between risk-sensitive foraging and predation-risk foraging.

Summary

After reading this chapter and before continuing with the material below, read the Chapter Summary on page 520. You will find related material on the following pages: population growth, 393-396; logistic equation, 393-396; carrying capacity, 393; intraspecific competition, 412-417; exploitative and interference competition, 398, 480.

Study Questions

1. Define predation, herbivory, cannibalism, and biophagy. (502)
2. What do the Lotka-Volterra and Nicholson-Bailey model of predation predict? (502-504)
3. How does the MacArthur-Rosenweig model modify those two models? (503-504).
4. What are the basic weaknesses of these models? (502-505)
5. How do laboratory studies support the predictions of these models? (505-509)
6. What is the functional response in predation? (508)
7. Distinguish between Type I, Type II, and Type III functional responses. (508-510)

8. Why is Type II functional response nonregulatory? (510)
9. What is the threshold of security? Compensatory predation? (510)
10. What is aggregative response? How does it affect intensity of predation? (510)
11. What is switching? (512)
12. What is a search image? How might it be related to switching? (512)
13. What is numerical response? (513)
14. What is optimal foraging? (514)
15. What is an optimal diet? What are the "decision rules" of an optimal diet? (515)
16. What is meant by foraging efficiency? What are some decision rules of optimal foraging? (515-516)
17. What is the marginal value theorem of foraging? (515)
18. How well do these rules hold up under natural conditions? (516-518)
19. What is the concept of satisficing and how does it relate to optimal foraging? (518)
20. What is risk-sensitive foraging? (519)
21. How does it relate to the expected energy budget rule? (519)
22. What is predation risk and how does it relate to foraging ? (519)

Key Terms and Phrases

parasitoidism	aggregative response
cannibalism	switching
herbivory	search image
biophagy	ratio-dependent predation theory
functional response	optimal foraging strategy
numerical response	marginal value theorem
Type I functional response	satisficing
Type II functional response	risk-sensitive foragingh
Type III functional response	predation
threshold of security	expected energy budget rule

Key Term Review

1. Certain wasps lay their eggs in the body of an insect. When the eggs hatch the wasp larvae feed on the insect, ultimately killing it. This combination of parasitism and predation is _____

2. When a predator eats proportionally more prey as the density of the prey increases, this is a _____ _____.

3. _____ occurs when a predator selects an alternate more abundant prey when its usual prey declines.

4. _____ occurs when the predator and the prey belong to the same species.

5. A rabbit feeding on the leaves of a shrub is an example of _____.

6. A direct _____ _____ takes place when the number of predators increase as the density of prey increases.

7. In a _____ _____ _____ _____, the number of prey taken per predator increases as the prey density increases up to the point of satiation.

8. A perceptual change in the ability of a predator to detect a cryptic prey is a _____ _____.

9. When predators tend to congregate in patches of high prey density, they are exhibiting an _____ _____.

10. When a population of a prey species has declined to a level at which it is not profitable for the predator to hunt it, the prey has reached its _____ ___ _____.

11. One organism feeding on another living organism is _____.

12. In a _____ _____ _____ _____, the number of prey taken per unit time decreases while the number of prey is still increasing.

13. An _____ _____ _____ for an animal provides a maximum net energy gain.

14. In a _____ _____ _____ _____, the number of prey taken per predator increases with increasing prey density and levels off when the ratio of prey taken to prey available declines.

15. How long a forager should profitably stay in a resource patch before it seeks another is given by the _____ _____ _____.

16. The fact that some animals will take some less than optimal food items and leave a patch before the food item is reduced to some minimal level is known as _____.

17. When an animal decides either to return to a patch that apparently offers a dependable supply of food or to visit a new patch where the availability and supply of food is unknown is called _____ _____.

18. A fish avoids a certain food-rich area of a pond because of the danger of being eaten there by a predator. This fish has to contend with _____ _____.

19. A rule of thumb for animals, be risk-prone if daily energy budget is negative; be risk-averse if it is positive, is called the _____ _____ _____ _____.

20. The _____ _____ _____ holds that both prey and predator equilibrium rise and fall with prey abundance.

Self Test

True and False

1. _____ A bird population moving into an area of insect abundance would represent a functional response.

2. _____ Solving the Lotka-Volterra predation equations results in oscillations in the predator-prey populations.

3. _____ A specialized predator would most likely seek alternative prey.

4. _____ Functional response plots prey taken against predator density.

5. _____ Type III functional response involves switching.

6. _____ Compensatory predation occurs when prey numbers increase above the thresholds of security and the surplus animals become more susceptible to predation through intraspecific competition.

7. _____ Functional response assumes that a predator population will eat proportionately more prey as the prey population increases.

8. _____ A self-sustaining predator-prey population requires an immigration of new prey to maintain itself.

9. _____ In a Type II functional response the number of prey taken increases as the prey density increases.

10. _____ When a new prey species appears in a predator's habitat it will become accepted prey in a short time.

11. _____ When a predator selects an alternate, more abundant prey over its usual prey, switching has taken place.

12. _____ Cannibalism is a type of intraspecific predation.

13. _____ Type II functional response often acts as a force that will stabilize a prey population.

14. _____ Gause showed that predators may overexploit their prey and die from starvation.

15. _____ Lotka and Volterra concluded that the growth rate of a predator population was influenced by the density of the prey population.

16. _____ Handling time by the predator is a dominant component of functional response.

17. _____ Aggregative response has little influence on predator-prey interactions.

18. _____ According to the marginal value theorem, a forager should stay in a food patch until the food resource is reduced below the average available at other patches.

19. _____ The "risk" in risk-sensitive foraging refers to the uncertainty a forager faces in locating food in a variable environment.

20. _____ When confronted with an energy shortfall, foragers seem to gamble on finding a new patch of food with the anticipation of a greater reward.

Matching

Select the appropriate terms to match with the statements below: (You will not use all of the terms.)

A. Switching
B. Type II functional response
C. Aggregative response
D. Type I functional response curve

E. Numerical response
F. Type III functional response
G. Search image

1. _____ Increases linearly as the number of prey taken per predator reaches the maximum it can possibly eat as the prey density increases.

2. _____ Predators move into an area of prey abundance.

3. _____ Predators increase because of an increasing and abundant food supply.

4. _____ Involves two or more types of prey.

5. _____ Associated with varying densities of one prey species.

1. In the functional response disk equation T represents the total time predator and prey are exposed. T is determined by the equation $T = T_s + T_h N_a$ where N_a is the number of prey or hosts killed. What does T_s and T_h involve? What variables are included in T_h?

2. Why is it difficult to test the optimal diet hypothesis under field conditions?

ANSWERS

Key Term Review

1. parasitoidism
2. functional response
3. switching
4. cannibalism
5. herbivory
6. numerical response
7. Type I functional response
8. search image
9. aggregative response
10. threshold of security
11. biophagy
12. Type II functional response
13. optimal foraging strategy
14. Type III functional response
15. marginal value theorem
16. satisficing
17. risk-sensitive foraging
18. predation risk
19. expected energy budget rule
20. ratio-dependent predation theory

Self Test

True and False

1. F	6. T	11. T	16. T	
2. T	7. T	12. T	17. F	
3. F	8. T	13. F	18. F	
4. F	9. T	14. T	19. T	
5. T	10. F	15. T	20. T	

Matching

1. D
2. C
3. E
4. F
5. B

Short Answer Questions

1. T_s is the time spent by the predator in search of prey. T_h is handling time. Handling time includes time spent pursuing prey, capturing prey, killing and processing prey, eating, and digestive pause, and the rest period between kills.

2. Many variables are involved in what an animal chooses to eat under natural conditions. Experimenters may have difficulty determining exactly what the animals are eating. They may have little information on the relative availability of different foods and what items are most profitable to take. The animals may choose foods that have lower relative availability and are less than optimal as an energy source. The forager may not have developed a strong search image for some available food items or initially lack experience in detecting or capturing the food item.

CHAPTER 24
PLANT-HERBIVORE SYSTEMS

Chapter Outline

Predation on Plants
    Effects on Plant Fitness
    Effects on Herbivore Fitness
Plant Defenses
    Mimicry
    Structural Defenses

Predator Satiation
Chemical Defenses
Herbivore Countermeasures
Functional Response of Herbivores
Models of Interaction
Summary

Learning Objectives

After completing this chapter you should be able to:

- Discuss the nature of predation on plants.
- Explain how herbivory affects plant fitness.
- Discuss how plants affect herbivore fitness.
- Describe the various plant defenses against herbivory.
- Explain how herbivores counteract plant defenses.
- Discuss the role of functional response in mammalian herbivory.
- Describe some models of plant herbivore interactions with some emphasis on plant herbivore cycles.

Summary

After reading this chapter and before continuing with the material below, read the Chapter Summary on page 533. You will find related material on the following pages: herbivores, 183; seed dispersal, 310, 590; logistic growth, 393-396; predator-prey systems, 505-508; Type II functional response, 508-509; search image, 512-515; optimal foraging, 515-519; defensive mutualism, 528, 587; herbivore-carnivore cycles, 543-545.

Study Questions

1. What two forms does predation on plants take? (522)
2. What effect does herbivory have on plant fitness? How do plants respond to defoliation? (522-524)
3. Why do grasses withstand herbivory much better than woody plants? (523)
4. In what way do plants affect herbivore fitness? (524)
5. How do the following mechanisms allow plants to defend themselves against herbivores: mimicry, structural defenses, and predator satiation? (525-529)
6. What two types of chemical resistance are employed by plants? (527-528)
7. What is the difference between apparent and unapparent plants? (528-529)

8. What are some ways in which herbivores breach a plant's chemical defenses? (528)

9. How does functional response relate to herbivory? How does functional response of mammalian herbivores differ from that of other types of predation? (529-530)

10. How do vegetation and herbivores interact in a predator-prey relationship? Discuss this interaction in relation to a vegetation-snowshoe hare cycle. How does chemical defense by plants enter the picture? (531-532)

Key Terms and Phrases

mimicry   predator satiation   apparent plants   unapparent plants

Key Term Review

1. _____ results when a palatable prey species resembles a distasteful species.

2. The timing of reproduction so that most of the offspring are produced in a very short time results in _____ _____.

3. Short-lived usually small perennials and annuals scattered in space and time that have highly toxic defenses are called _____ _____.

4. Long-lived, usually woody plants, that possess dosage-dependent toxic defenses such as tannins and resins are _____ _____.

Self Test

True and False

1. _____ For a herbivore, the quality of food is more important than the quantity available.

2. _____ Secondary substances in plants can affect reproductive success in herbivores.

3. _____ Many herbivores, especially insects, are able to detoxify chemical defenses of plants.

4. _____ Predator satiation is a major defensive mechanism in plants.

5. _____ Plants probably evolved secondary metabolites for predator defense.

6. _____ Plants have a distinct advantage in predator-prey relationships.

7. _____ Structural defenses are mostly presumed.

8. _____ The effect of structural defenses is to increase handling time by herbivores.

9. _____ Plant toxins provide a defense against a full suite of insect enemies.

10. _____ Seed predators seek seeds scattered some distance from the parent plant.

11. _____ Some feeding specialists are able to eat very toxic plants.

12. _____ Some secondary metabolites in plants act as contact poisons.

13. _____ Predator satiation is most prevalent in those plants possessing strong structural defenses.

14. _____ Seed predation is true plant predation.

15. _____ Removal of foliage has little effect on a plant's competitive position in a stand.

16. _____ Mimicry is widely evolved in plants.

17. _____ Moderate grazing can have a stimulating effect on plants, but at some cost in vigor.

18. _____ The impact of seed predation is easy to assess.

Matching

Associate each of the following statements with apparent (**A**) or unapparent plants (**U**).

1. _____ Chemical defenses are not easily mobilized.

2. _____ Defenses are highly toxic.

3. _____ Toxins are effective at low concentrations.

4. _____ Mostly large woody plants.

5. _____ Toxic reserves are concentrated near the surface of the leaf.

6. _____ Reduces the rate herbivore assimilation.

7. _____ Toxins are readily transported to site of attack.

8. _____ Substances can be synthesized at relatively low cost.

Associate each of the strategies given below with the type of plant defense.

A. Chemical   B. Predator satiation   C. Structural defense   D. Mimicry

9. _____ Scatter seeds away from parent plant.

10. _____ Accumulation of secondary metabolites.

11. _____ Leaf stipules resemble the size and shape of insect eggs.

12. _____ Thick, hard seed coats.

13. _____ Prickles on leaves.

14. _____ Synchronized seed production.

15. _____ Causes illness or reduce digestibility in herbivores.

Short Answer Questions

1. Why is predation on plants by grazing herbivores more like parasitism than predation?

2. What affects the quality of plant food for herbivores?

3. What are the two general types of chemical defenses in plants?

4. How does functional response involving grazing herbivores differ from functional response involving carnivory.

ANSWERS

Key Term Review

1. mimicry   2. predator satiation   3. unapparent plants   4. apparent plants

Self Test

True and False   *Matching*

1. T	7. T	13. F	1. A	6. A	11. D
2. T	8. T	14. T	2. U	7. U	12. C
3. F	9. T	15. F	3. U	8. U	13. C
4. T	10. F	16. F	4. A	9. B	14. B
5. F	11. T	17. T	5. A	10. A	15. A
6. F	12. T	18. F			

Short Answer Questions

1. Predation on animals and on seeds kills the organism. With exceptions, such as spruce budworms feeding on conifers, herbivores withdraw nutrients from plants in the form of green parts and sap. They do not kill the plant, just as parasites generally do not kill their hosts, although they affect the vigor of the plant, just as parasites can affect the vigor of the host.

2. Quality of plant food is influenced by age. Young growing tissues are more nutritious, higher in nitrogen, and more palatable and digestible than mature tissues. Low quality foods are generally fibrous, low in nutrients, and hard to digest. Also affecting the quality of food are chemical defenses that lower digestibility.

3. In one type of chemical defense the plant accumulates and changes toxins at the wound site. This method comes into play when the plant leaf or structure is attacked. The other type is based on the presence of substances, such as tannins, that discourage the animal from feeding on the plant.

4. The components of functional response involving grazing herbivores is more complex than functional responses involving animal predators and prey. Because of the nature of their growth, plants are more subject to attack and to repeated attacks. The distribution of plants, which involves hidden, loosely scattered, and concentrated patches, together with the variety of food available affect the availability and choice of food. The attack rate of herbivores on plants is measured in bite size and how fast the grazer chews. Both of these are influenced by the size of the mouth and the nature of the plant. The amount of plant tissue taken in each bite declines as the intake of plant tissue declines. During feeding, grazing herbivores can chew food as they move to new patches. As a result, search time and handling time overlap.

CHAPTER 25
HERBIVORE-CARNIVORE SYSTEMS

Chapter Outline

Prey Defense
  Chemical Defense
  Warning Coloration and Mimicry
  Cryptic Coloration
  Armor and Weapons
  Behavioral Defenses
  Predator Satiation
Predator Offense
  Hunting Tactics
  Cryptic Coloration and Mimicry
  Adaptations for Hunting

Cannibalism
Intraguild Predation
Predator-Prey Cycles
Regulation
Exploitation by Humans
  Sustained Yield
  Problems with Management
Summary

Learning Objectives

After completing this chapter you should be able to:
- Discuss the array of defensive tactics prey have evolved to escape predation.
- Describe some tactics predators have evolved to capture prey.
- Discuss the nature and role of cannibalism in populations.
- Explain intraguild predation.
- Discuss the forces behind predator-prey cycles.
- Discuss the role of predation in population regulation.
- Tell why human exploitation of populations is often not related to the density of prey.
- Explain the meaning of sustained yield and how it differs between *r*-selected and *K*-selected species
- Argue that economics rather than ecology controls the exploitation of populations.
- Point out the signs of overexploitation of populations.

Summary

After reading this chapter and before continuing with the material below, read the Chapter Summary on page 556. You will find related material on the following pages: animal nutrition, 124; logistic growth, 393-396; population regulation, 396-402; population fluctuations, 403-406; altruistic behavior, 463; plant-herbivore interactions, 522; plant mimicry, 525; chemical defense in plants, 527-529; herbivore-vegetation cycles, 531-533; parasitism, 558-573.

Study Questions

1. What are the variety of defenses employed by prey species against predation?
2. What is mimicry? What is the model? The mimic?
3. What are some behavioral forms of defense against predation?
4. How can predator satiation reduce the impact of predation?
5. What are some predator countermeasures against prey defenses?
6. What is intraguild predation? How does it influence predator-prey relationships, especially when new species are introduced into an ecosystem? (542-543)
7. What seems to trigger cannibalism in a species? What are some of the selective advantages? Disadvantages?
8. Why is cannibalism disadvantageous from an evolutionary standpoint?
9. Can predators regulate prey species; if so under what conditions?
10. What members of a prey species are most vulnerable to predation?
11. What is the relationship between predator and prey cycles? What seems to drive the 10-year cycle?
12. What is sustained yield as it relates to human exploitation of natural populations?
13. What are some major points to be considered in the harvesting of natural populations such as a fishery?
14. What are some signs of overexploitation of a population?
15. What is the relationship between economics and the management of exploited populations?

Key Terms and Phrases

warning coloration	yield	maximum sustained yield
cryptic coloration	biomass yield	optimum sustained yield
aggressive mimicry	standing crop	fixed quota
cannibalism	productivity	harvest effort
intraguild predation	sustained yield	dynamic pool model
attack-abatement effect	Batesian mimicry	Mullerian mimicry

Key Term Review

1. _____ occurs when predator and the prey are the same species.

2. Patterns, shapes, and color that make prey less visible is _____ _____.

3. Predators that deceive prey by resembling the prey, are employing _____ _____.

4. Animals that possess toxicity and possess bold coloration discourage predators by _____ _____.

5. The killing and eating of one species by another that use similar resources and thus are potential competitors is _____ _____.

6. Prey living in a group reduces an individual's chance of being taken and deter attacks by predators. These advantages of group living create an _____ _____.

7. Individuals or biomass removed from an exploited population is _____.

8. _____ _____ is the weight of individuals removed from the population.

9. The difference between the biomass left in a population after harvesting and the biomass present before harvesting is _____.

10. The amount of biomass or numbers in a population at the time it is measured is the _____ _____.

11. _____ _____ is achieved when the yield per unit time equals productivity per unit time.

12. The level of sustained yield at which a population declines if exceeded is the _____ _____ _____. A yield that takes into consideration species interactions and other factors is the _____ _____ _____.

13. Removing a certain percentage of the population at each harvest period based on MSY is the _____ _____ approach to harvesting.

14. An approach to harvesting that involves increasing or decreasing the number of animals taken by controlling the number of hunters, boats in a fishing fleet, or season length is _____ _____.

15. An approach to harvesting a population, particularly fish, that assumes 1) a constant natural mortality rate that is independent of population density and 2) the animals removed replace natural mortality is the _____ _____ _____.

16. In _____ _____ both models and mimics are unpalatable; in _____ _____ the model is unpalatable and the mimic is palatable.

Self Test

True and False

1. _____ Vertebrate population cycles appear to be driven by predation.

2. _____ When humans first harvest a population, they tend to take the smaller, younger individuals first.

3. _____ Economics rather than biology has dictated the rates of harvest of fisheries.

4. _____ There is only one level of exploitation for maximum sustained yield.

5. _____ The rate of harvest should not exceed the rate of increase the population would experience if it were not being harvested.

6. _____ Reproductive synchrony functions best against generalist predators.

7. _____ Cannibalism has only a minimal effect on the dynamics of a population.

8. _____ The resemblance of an edible species to an inedible one is Mullerian mimicry.

9. _____ Mimicry can be most advantageous to the model if the mimics are abundant relative to the number of models.

10. _____ Some insects gain active components for chemical defense from toxic plants on which they feed.

11. _____ The most energetically expensive method of hunting for a predator is stalking.

12. _____ A specialized predator cannot regulate an ungulate prey.

13. _____ Among insects, over half of the cannibalistic species are herbivorous.

14. _____ Cannibalism provides little selective advantage to survivors.

15. _____ The young of interacting species are most vulnerable to intraguild predation.

16. _____ There is evidence that sunspot cycles can synchronize the nine-to-ten year snowshoe hare cycle across boreal regions.

17. _____ Malnutrition appears to trigger and predation to drive the nine-to-ten year cycles of hare and lynx.

18. _____ Introduction of a new predatory species into freshwater ecosystems can intensify intraguild predation.

19. _____ A diversity of habitats and alternate prey inhibit cycles south of the boreal region.

20. _____ Northern rodents exhibit a three-to-four year cycles.

Matching

Match the statement with the appropriate term (You will not use all the terms.)

A. Standing crop
B. K/2
C Four
D. Contest-type competition

E. Maximum sustained yield
F. Scramble competition
G. I. Sustained yield

H. Harvest rate
I.   Rate of increase
J. Two
K. K/4

1. _____ $H = r$.

2. _____ Population declines if yield exceeds this level.

3. _____ Biomass present in population at time it is measured.

4. _____ Number of levels of density from which the same sustained yield can be obtained.

5. _____ A population must be reduced below this level to obtain a croppable surplus.

6. _____ Rate of exploitation can be relatively high in this type of population.

Labeling

1. On the reproductive diagram below plot and label the following points: maximum sustained yield (MS1) , and two points where sustained yield S_1 = sustained yield S_2.

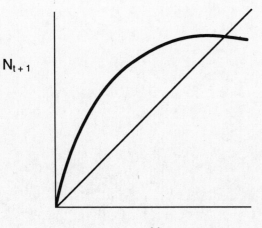

Short Answer Questions

1. What can trigger cannibalism in a population?

2. What is intraguild predation and what is its significance?

3. What are signs of overexploitation of a population?

ANSWERS

Key Term Review

1. cannibalism
2. cryptic coloration
3. aggressive mimicry
4. warning coloration
5. intraguild predation
6. attack-abatement effect
7. yield
8. biomass yield
9. productivity
10. standing crop
11. sustained yieldt
12. maximun sustained yield
    optimal sustained yield
13. fixed quota
14. harvest effort
15. dynamic pool model
16. Mullerian mimicry
    Batesian mimicry

Self Test

True and False

1. T	8. F	15. T
2. F	9. F	16. T
3. T	10. T	17. T
4. T	11. F	18. T
5. T	12. F	19. T
6. F	13. T	20. F
7. F	14. F	

Matching

1. H
2. E
3. A
4. J
5. B
6. F

Labeling

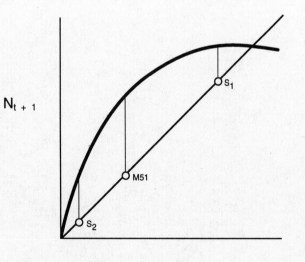

Short Answer Questions

1. Cannibalism probably starts when hunger triggers search behavior in the animal, lowers its threshold of attack, and increases its search area until it comes in contact with the first vulnerable prey of the same species. Stress and high populations can promote cannibalism even though food may be adequate as well as the presence of vulnerable individuals such as young and runts.

2. Intraguild predation involves elements of competition and predation. In its simplest form intraguild predation involves two species that share the same resources but one is a predator. One prey species may share the same resource as the young of the predaceous species. By preying on the competing species, the predator increases the resource base for its own young and reduces the population of its intraguild competitor.

3. Changes that warn of overexploitation of a population include: a) a decreased catch per unit effort; b) a decreasing proportion of females carrying young; c) failure to increase young after harvesting; and d) a high proportion of young, nonreproducing individuals.

CHAPTER 26
PARASITISM

Chapter Outline

Characteristics of Parasites
Hosts as Habitats
Life Cycles
   Direct Transmission
   Indirect Transmission
   Dynamics of Transmission
Host Responses
   Biochemical Responses
   Abnormal Growth
   Behavioral Changes
   Mate Selection

Population Dynamics
   Models of Parasitism
   Parasitism as a Regulatory Mechanism
   Parasite Population Dynamics
Evolutionary Responses
Social Parasitism
   Brood Parasitism
   Kleptoparasitism
Summary

Learning Objectives

After completing this chapter you should be able to:
- Distinguish between microparasites and macroparasites.
- Discuss hosts as habitats for parasites.
- Distinguish between direct and indirect transmission of parasites.
- Discuss factors involved in transmission of parasites.
- Explain how plant and animal hosts respond to parasitic infections.
- Describe a population model of parasitism.
- Discuss parasitism as a possible regulatory mechanism of host populations.
- Describe some unique aspects of parasite population dynamics.
- Discuss the evolutionary responses between parasites and hosts.
- Discuss the significance of social parasitism, especially brood parasitism.

Summary

After reading this chapter and before continuing with the material below, read the Chapter Summary on page 577. You will find related material on the folllowing pages: population growth, 392-396; intraspecific competition, 412-414; sexual selection, 430-436; natural selection, 451-463; interspecific competition, 480-513; predator-prey relationships, 536-547; coevolution, 582; defensive mutualism, 587; parasites and community structure, 613-624.

Study Questions

1. What is parasitism? Distinguish between a parasite and a parasitoid. (558)
2. What is disease? (558)
3. Distinguish between a macroparasite and a microparasite. What groups are included in each? What type of immune responses does each stimulate? (558-559)
4. What is a definitive host? An intermediate host? (560)
5. Distinguish between direct transmission and indirect transmission. Give an example of each. (560-564)
6. What is the relationship between population density and predation in the transmission of parasites? (564-566)
7. Distinguish between an epizootic and an epidemic? (565)
8. What are several host responses to parasitism? (566-568)
9. How might parasites inhibit sexual reproduction? Mate selection? (567-568)
10. Can parasites regulate a host population? If so, how? (569)
11. What factors are involved in the growth dynamics of a parasite population? (571-572)
12. How do parasites and their hosts develop a "mutual tolerance"? (572-573)'
13. What is social parasitism? What forms does it take? (573-577)

Key Terms and Phrases

microparasites	definitive host	holoparasites	epizootic
macroparasites	intermediate host	hemiparasites	epidemic
kleptoparasitism	vector	parasitism	brood parasitism
parasitoids	disease	social parasitism	

Key Term Review

1. The _____ _____ is the one in which the parasite becomes an adult.

2. _____ is an interspecific interaction in which two organisms live together and one gains its nourishment from the tissues of the other.

3. The _____ _____ is the one that harbors the developmental phase of a parasite.

4. Organisms that carry or transmit a parasite from one host to another are called _____.

5. Parasites that essentially act as predators are _____.

6. Plants that lack chlorophyll and draw water, nutrients, and carbon from the roots of other plants are _____.

7. Plants that are photosynthetic but draw water and nutrients from the host plant are _____.

8. Parasites characterized by a very small size, a short generation time, and multiply rapidly within their host are _____.

9. The rapid spread of viral and bacterial diseases in dense populations is called an _____ among humans and an _____ among animals.

10. _____ are large in size, have a relatively long generation time, and rarely multiply directly in the host.

11. The foisting of the incubation of eggs or care of young on foster parents is _____ _____.

12. A bald eagle forcing an osprey to give up its prey is an example of _____.

13. Any conditions that deviates from normal well being is _____.

14. When one organism is parasitically dependent on the social organization of another it is experiencing _____ _____.

Self Test

True and False

1. _____ The adult stage of a macroparasite, such as tapeworm, inhabits the definitive host.

2. _____ The rapid spread of rabies through a fox population is an epizootic.

3. _____ All parasites cause disease.

4. _____ The cowbird, which lays its eggs in the nests of other birds, is a temporary obligate social parasite.

5. _____ The transmission of a parasite depends heavily on the density of host populations.

6. _____ Microparasites multiply in their host.

7. _____ Rabies will confer a long-term immunity on its host.

8. _____ Microparasites are the vectors for many diseases.

9. _____ Macroparasites have a short generation time.

10. _____ Direct transmission of plant viral diseases commonly employs insect vectors.

11. _____ Immunity of a host may eliminate a parasite.

12. _____ A large value of R_o in a microparasite population indicates a high rate of infection.

13. _____ A parasite that kills its host kills itself.

14. _____ A high infestation of parasites in a host can result in a high growth rate of the parasite.

15. _____ The transmission threshold $R_o = 1$ must be crossed if a disease is to spread.

16. _____ Low virulence encourages a long duration of a parasite in a host.

17. _____ Plant galls are an example of a growth response to parasitism.

18. _____ Natural selection favors a more virulent form of parasitism.

Matching

Consider two type of parasites, a viral microparasite: **A** *Distemper* and an internal macroparasite parasite: **B** *lungworm*. Below are some characteristics of parasitic infections. Associate each of the following with the parasites named above by indicating an **A** or **B** for the appropriate characteristic.

1. _____ Direct transmission.

2. _____ Long-term infection.

3. _____ Acquired immunity.

4. _____ Requires more than one type of host.

5. _____ May function as a population regulatory mechanism.

6. _____ Direct multiplication in host.

Short Answer Questions

1. Ecologically parasites can be divided into two groups. Name the two groups and characterize each according to the following:

Group	Size	Length of Infection	Common type of transmission

216

2. Distinguish between direct transmission and indirect transmission of a parasite and give an example of each for a plant parasite and an animal parasite.

3. What problems do parasites face maintaining their own survival and population growth?

4. Explain how the population growth of a parasite is closely tied to the population growth rate of its host?

5. Contrast obligatory brood parasitism with temporary brood parasitism and give an example of each.

ANSWERS

Key Term Review

1. definitive host	6. holoparasites	10. macroparasites
2. parasitism	7. hemiparasites	11. brood parasitism
3. intermediate host	8. microparasites	12. kleptoparasitism
4. vector	9. epidemic	13. disease
5. parasitoids	epizootic	14, social parasitism

Self Test

True and False

1. T	7. F	13. T
2. T	8. F	14. F
3. F	9. F	15. T
4. T	10. T	16. T
5. T	11. F	17. T
6. T	12. T	18. F

Matching

1. A
2. B
3. A
4. B
5. A
6. A

Short Answer Questions

1. The groups and characteristics are:

Group	Size	Length of Infection	Common type of transmission
Microparasites	Very small	Short	Direct
Macroparasites	Relatively large	Long duration	Indirect; ectoparasites, direct

2. In direct transmission the parasite spreads by contact between hosts. An example is the common cold and AID viruses. Some direct transmissions require a vector, such as malarial parasites carried by mosquitoes and certain plant diseases carried by leaf hoppers. Many macroparasites of plants and animals are spread by direct contact. Examples are mange and fleas in animals and fungal infections in plants. Indirect transmission involves an intermediate host or hosts in which the infective stages develop and though which the parasites are transmitted to the definitive host. Examples are lungworm and tapeworm infections in animals and rusts in plants.

3. Parasites face major problems unique to them because their survival and growth rates are closely related to their living hosts. These include, among others, available living space within a host, how the parasites are distributed among the hosts, intraspecific competition within the host, longevity of the hosts, availability of intermediate hosts where needed, and clumped distribution of hosts that influences probability of a parasite finding a host.

4. For a parasite the host is both food and habitat. Thus the fate of a parasitic population is related to the density of the host population and the distribution of individual parasites among the host population. A host can support only so many adult parasites before the parasites experience intraspecific competition or kill their host. If the parasite has infective stages, functioning as individual populations, that require one or more intermediate hosts, then population growth is influenced by the distribution and size of those hosts. A parasitic population cannot grow until it has reached some threshold size within the host and it cannot spread if another host is not available.

5. Temporary obligatory brood parasitism involves those species that have lost the ability to parent their own young. Their reproductive success depends upon the successful rearing of their young by a foster parent of another species. Fostering offspring of brood parasites reduces the fitness of the foster parent because the young parasite is reared at the expense of the foster parent's own young. An example is the European cuckoo. Facultative brood parasitism involves a number of bird species that that nest and rear their own young but deposit eggs in the nest of a related species or their own species. Such parasitism may or may not reduce the fitness both of the host and the parasite.

CHAPTER 27
MUTUALISM

Learning Objectives

After completing this chapter you should be able to:
- Define coevolution.
- Describe the different types of mutualism.
- Discuss the relationship between pollination and mutualism.
- Explain the role of mutualism in seed dispersal.
- Discuss the evolution of mutualism.
- Discuss the difficulties of modeling mutualistic relations between species.

Summary

After completing this chapter and before continuing with the material below, read the Chapter Summary on page 594. You will find related material on the following pages: adaptation, 3-31; nutrient cycling, 116-120; rumen metabolism, 183; nitrogen fixation, 206-207; function of temperate forests, 277-280; function of tropical forests, 285-287; coral reefs, 344; natural selection, 461-463; interspecific competition models, 480-484; predator-prey interactions, 536-547; parasite-host relationships, 560-568.

Study Questions

1. What is coevolution? (580)
2. Define mutualism? (580)
3. Distinguish between symbiosis, symbiotic mutualism, and nonsymbiotic mutualism. (580)
4 Distinguish between obligate symbiotic mutualism, obligate nonsymbiotic mutualism, facultative mutualism, defensive mutualism, and indirect mutualism. (581-585)
5. What are mycorrhizal mutualisms? What is the importance of these mutualisms? Why are they obligate? (582-583)
6. In what ways are seed dispersal and pollination forms of facultative mutualism? (585-591)
7. How does defensive mutualism function? (585)

8. How might have mutualistic relationships evolved? (591)
9. Why is it difficult to model population effects of mutualism? (592-594)

Key Terms and Phrases

coevolution
diffuse coevolution
mutualism
symbiosis
symbiotic mutualism
diffuse coevolution

nonsymbiotic mutualism
ectomycorrhizae
endomycorrhizae
obligate nonsymbiotic mutualism
obligate symbiotic mutualism

defensive mutualism
indirect mutualism
elaiophores
myrmecochores
facultative mutualism

Key Term Review

1. Plants that have an ant-attracting food body on their seed coat are called
   _____.

2. When certain traits of one species evolve in response to the traits of another
   species, the process is called _____.

3. The relationship between two organisms living in close physical association from
   which one or both derive a benefit and at least one of the two species cannot live
   independently is _____.

4. In _____ _____ both members of the pair benefit
   each other, but they live independently.

5. One form of a nutritional relationship between fungi and plant roots that involves a
   well-developed fungal sheet or mantle around the root is _____.

6. One form of mutualism in which the relationship is so permanent that the distinction
   between the two organisms becomes blurred is _____
   _____ _____.

7. Mutualism that is nonobligatory and opportunistic is _____
   _____.

8. _____ infect roots by penetrating the cells of the host to form a
   finely branched network.

9. Fungal endophytes that infect the leaf tissues of grasses making the grass
   unpalatable to grazing herbivores are examples of _____
   _____.

10. _____ _____ occurs in competitive situations in which the
    interactions between two species benefits the well-being of a third species.

11. Oil-secreting organs on flowers and seeds that attract some pollinators and seed
    dispersers are _____.

12. _____ _____ occurs when the adaptive responses are spread over many interacting species.

Self Test

True and False

1. _____ Mutualism may have evolved from commensal relationships.

2. _____ Mutualisms are best appreciated at the population level.

3. _____ Obligate nonsymbiotic mutualistic relationships cost both mutualists.

4. _____ Nectar and oils are of no value to plants except to attract pollinators.

5. _____ Plants prevent the eating of unripe fruit by cryptic or concealing coloration.

6. _____ Generalist frugivores are obligates of the fruits they eat.

7. _____ Orchids offer rich rewards to their pollinators.

8. _____ Tropical orchids depend entirely upon certain bees for pollination.

9. _____ Lack of appropriate mycorrhizae in the soil can inhibit the colonization of a site by plants.

10. _____ Zooanthellae are photosynthetic dinoflagellates living in coraline anthozoans.

11. _____ Mycorrhizae are especially important in nutrient-rich soils.

12. _____ In the algal-fungal relationship in lichens, the fungus may be parasitic on the lichen.

Matching

Categorize the following mutualistic relationships by associating the term with the appropriate statement:

A. Obligate symbiotic mutualism
B. Facultative mutualism

C. Defensive mutualism
D. Obligate nonsymbiotic mutualism

1. _____ Fungi growing in plant tissue make the host grass unpalatable to grazing herbivores.

2. _____ Bees carry pollen from one flower to another.

3. _____ Mycorrhizae growing on the roots of pine trees.

4. _____ Yucca moth larvae feed on seeds of yucca; yucca in turn depends upon yucca moths for pollination.

5. _____ Dogwoods and wild grape depend upon birds to disperse their seeds.

Short Answer *Questions*

A. Oaks and pines require certain fungi known as mycorrhizae to parasitize their roots. These fungi penetrate the root tissues and extend into the surrounding soil where they function as additional rootlets drawing nutrients from the soil. The fungi must live on the tree roots; the oaks and pines grow poorly without the aid of mycorrhizae.

B. In the tropical rain forest certain species of ants cultivate a fungus in their nests for food. The fungus cannot live without cultivation by ants; the ants cannot live without the fungus as food.

1. Which of the above cases represents obligate symbiotic mutualism and why?

2. Which of the above is nonsymbiotic obligate mutualism and why?

3. Many one-to-one relationships between two species--predator and prey and mutualists--appear to have evolved over a long period of time. Argue that this apparent relationship may not be one that coevolved.

4. Why do some plants evolve mechanisms to attract specific pollinators, whereas pollinators themselves are generalists?

5. Most fruiting plants depend upon animals to disperse their seeds by eating the fruits. How do plants attract frugivores to their fruits? What two alternative dispersal means do plants employ?

Concept Application Exercise/ Chapters 22-27.

Read the following narrative carefully. Questions testing your ability to apply concepts to the situations described in the narrative will appear at the end.

Consider a countryside of agricultural land,. abandoned farms, and woods. Living in this countryside are numerous species of wildlife including bobwhite quail, red fox, bluebirds, gray squirrels, and others.

In the woods lives a sizable population of gray squirrels, attracted there by stands of old, mature oaks. Some years the white oaks have a large crop of acorns; other years red oaks are prolific. In the years of heavy acorn production gray squirrel, deer, and turkey cannot eat all the acorns. Red oak acorns, high in tannin, are eaten sparingly by gray squirrels.

Adjacent to the woods is a recently abandoned field with patches of open soil used by a pair of killdeer as a nesting site. When disturbed, the incubating killdeer leaves the nest quickly while dragging its wings and spreading its bright orange tail. The eggs, flecked and mottled with black, brown, and white, and placed in a shallow depression on the bare ground, so match the ground they are hard to see.

The bobwhite quail had been fairly stable over the past few years. But recently a landowner cleaned out most of the hedgerows between hay and crop fields, greatly reducing winter cover. Last winter a Cooper's hawk appeared and over the winter eliminated several flocks of quail trying to live in narrow border of shrubs along the woods. In previous years the Cooper's hawk had a difficult time capturing quail. It pursued the quail through the thickets, but the quail usually escaped in dense ground cover. With the cover gone, the Cooper's hawk could capture quail easily and fly into the woods to eat its prey.

Living in the countryside are red foxes. During the spring the foxes take some quail, but mostly they live on frogs and young rabbits in the spring and concentrates on grasshoppers, mice, and berries in late summer, and mice and rabbits in winter. While feeding on mice, the foxes picked up the tapeworm Taenia taeniaformis *by eating meadow mice that carried the immature stages of the tapeworm in their liver.*

Frequenting the farmsteads and fields are number of starlings. A cavity nester, the starlings drive out other cavity nesters, particularly flickers and bluebirds. and occupy these limited nesting sites. The starlings nest, but some bluebirds cannot because they cannot find nesting cavities.

Starlings eat corn borers and other insect larvae. One particular area in the region experienced an outbreak of corn borers. The abundance of these insects in corn fields attracted a large number of starlings. The birds fed heavily on the borers. The number of larvae taken by the starlings increased for a time, but eventually the starlings took a more or less constant number of prey.

Some of the old fields were colonized by annual weeds including bristly foxtail, Indian mallow, and smartweed. Bristly foxtail has shallow surface roots. Indian mallow has a taproot reaching intermediate depths in the soil. Smartweed has a root system well developed below the rooting zone of the other two species. Thus each species uses a different zone in the soil for moisture and nutrients.

Patches of goldenrod grow in the old fields, apparently colonizing places where grass cover had died out. Feeding on the goldenrod one summer were numbers of the beetle Galerucella americana. *Although heavily preyed upon by birds living in the area, the insect was able to maintain a sufficiently large population to lay eggs and carry the population through the winter to the next spring and summer.*

Back in the woods, a number of trees supported growths of mistletoe in their crowns. A flock of overwintering cedar waxwings fed on the whitish fruits. The seeds, covered by a sticky coating, passed through the birds' digestive tracts unharmed and stuck to the twigs and limbs of the host trees, where some of them germinated.

Select the concepts listed below as matching items best fitting the following statements based on the narrative. (You will not use all of the matching items.)

A. Handling time	K. Switching	T. Generalist
B. Narrow	L. Specialist	U. Wide
C. Refuge	M. Stalk	V. Chemical defense
D. Cryptic coloration	N. Search time	W. Indirect mutualism
E. Patchy environment	N. Mimicry	X. Functional response
F. Interspecific competition	O. Parasitism	Z. Obligate nonsymbiotic mutualism
G. Numerical response	P. Immigration	AA. Warning coloration
H. Pursuit	Q. Handling time	BB. Resource partitioning
I. Distraction display	R. Definitive host	CC. Indirect transmission
J. Direct transmission	S. Satiation	DD. Intermediate host

1. _____ Predator defense by red oaks.

2. _____ Predator defense by white oaks (and incidentally by red oak as well).

3. _____ Probable reason for the goldenrod beetle to maintain its population in face of heavy predation by birds.

4. _____ Evolved response of the three plants to competition for water and nutrients.

5. _____ Defense against nest predation employed by the killdeer.

6. _____ Influx of starlings into the cornfields because of an abundance of prey.

7. _____ As the number of corn borers emerged, more of them were taken by starlings, initially at least.

8. _____ Relationship between starling and bluebird at nesting time.

9. _____ Predatory behavior exhibited by the red fox.

10. _____ Niche width of the Cooper's hawk.

11. _____ Component of predation involved in pursuit, capture, and eating of prey by the hawk.

12 _____ From niche theory, type of predator represented by the red fox.

13. _____ In terms of predation theory, what function did the shrubby hedgerows serve?

14. _____ Egg or nest concealment by killdeer.

15. _____ Relationship of the mistletoe to the oak tree.

16. _____ Type of hunting method used by the Cooper's hawk.

17. _____ Relationship between mistletoe and cedar waxwing.

18. _____ Type of host represented by the meadow mouse.

19. _____ Type of transmission used by the parasitic tapeworm.

ANSWERS

Key Term Review

1. myrmecochores
2. coevolution
3. symbiosis
4. nonsymbiotic mutualism
5. ectomycorrhizae
6. obligate symbiotic mutualism
7. facultative mutualism
8. endomycorrhizae
9. defensive mutualism
10. indirect mutualism
11. elaiophores
12. diffuse coevolution

Self Test

True and False

1. T
2. F
3. T
4. T
5. T
6. F
7. F
8. T
9. T
10. T
11. F
12. T

Matching

1. C
2. B
3. A
4. D
5. B

Short Answer Questions

1. Mycorrhizae on oaks and pines is an obligate symbiotic mutualism. The relationship is symbiotic because mycorrhizae become part of the physical structure of the rootlets. It is obligate because both the mycorrhizae and the trees are dependent upon each other, the trees for nutrients and the mycorrhizae for energy.

2. The ant-fungal relationship is nonsymbiotic obligate mutualism because the ants and the fungi are physically separate and lead independent lives. It is obligate because ants and fungi are wholly dependent upon each other for their survival.

3. Many apparent one-to-one relationships may have evolved between two species in different places over time. When one species invades a new habitat, it may be able to use the trait acquired to fit the new situation. The new observed interaction suggests a long-term coevolution. On a larger scale some traits evolved in several species in one taxon in response to traits in another species. Plants adapt to a new spectrum of pollinators and the pollinators adapt to a new range of plants. This interaction is called diffuse coevolution.

4. Many plants flower simultaneously and compete for pollinators. They need to attract specific pollinators to ensure pollen transfer to individuals of their own species and to avoid wastage. Flowering of various species, however, is time limited. Pollinators can act as specialists only temporarily. To survive they depend upon a succession of nectar-producing flowers through the season. Energetically, pollinators cannot afford to specialize; they must remain generalists.

5. Fruiting trees depend upon fruit-eating animals to disperse their seeds. When their fruits are ripe, plants coax frugivores with attractive odors, conspicuous colors, palatable textures, and high sugar content. Plants employ two alternative methods. One approach is opportunistic. Plants evolved fruits that attract an array of generalist fruit eaters that will disperse seeds through the gut. These plants opt for quantity dispersal of seeds, with the chance that a few will settle in suitable places for germination and growth. This type of dispersal is common in temperate regions. The second alternative is to depend upon a small number of birds and mammals that are exclusively frugivorous and temporary specialists. They disperse the fruit of different species through the year. The seeds are not widely scattered, but usually dropped beneath or some distance away from the plants. This alternative is typical of tropical regions.

Concept Application Test

1. V	8. F	15. O
2. S	9. K	16. H
3. P	10. B	17. Z
4. BB	11. Q	18. DD
5. I	12. T	19. CC
6. G	13. C	
7. X	14. D	

PART 6

THE COMMUNITY

CHAPTER 28
COMMUNITY STRUCTURE

Chapter Outline

The Community Defined
Physical Structure
   Life Forms
   Vertical Stratification
   Horizontal Structure
Biological Structure
   Species Dominance
   Species Diversity
   Species Abundance
Edge Communities
Island Communities
   Island Biogeography Theory
   Habitat Fragmentation

   Corridors
   Assembly Rules
Population Interactions
   Influence of Competition
   Influence of Predation
   Effects of Parasites and Diseases
   Effects of Mutualism
Community Patterns
   Organismic vs Individualistic
   Gradients and Continua
Classification Systems
Summary

Learning Objectives

After completing this chapter you should be able to:

- Define the community.
- Describe vertical stratification and horizontal structure of communities.
- Discuss the factors that influence local, regional, and global diversity.
- Explain the concepts of species dominance, species diversity, and species abundance.
- Describe the nature of edge and ecotone.
- Explain the island biogeography theory.
- Discuss habitat fragmentation and its relation to the island biogeography theory.
- Describe the role of corridors relative to habitat fragmentation.
- Discuss the influence of competition, predation, parasitism, and mutualism on community structure.
- Contrast the organismic and individualistic concepts of the community.
- Distinguish between a gradient and a continuum.
- Discuss various approaches to community classification.

Summary

After reading this chapter and before continuing with the material below, read the Chapter Summary on pages 629-631. You will find additional material on the following pages: adaptation, 30-31; energy flow and food webs, 189-193; grassland structure, 231-233; forest structure, 268, 276-277, 283-285; lakes and ponds, 294-297; effective population size, 472-474; viable populations, 475; vegetation-herbivore interactions, 522-524; mycorrhizae, 528, 584; predator-prey interactions, 536-547; population effects of parasitism, 569-572; mutualism, 582-596; moisture gradients, 627; succession, 656-667. Appendix, 689-696, 722-728.

Study Questions

1. Distinguish between a community and a guild. (598)
2. Describe the Raunkiaer life forms. (598-599)
3. What is vertical stratification of a community? Why is it important? (599-601)
4. What is involved in the horizontal structure of a community? How does it relate to a patchy environment? (601)
5. How does the physical structure of a community relate to its biological structure? (602)
6. Distinguish between species dominance, species richness, and species diversity. How is the latter related to dominance? (602-604)
7. What are the methods of determining dominance? (602, Appendix 689-696)
8. Distinguish between local diversity and global diversity.
9. What hypotheses have been proposed to explain latitudinal and altitudinal species diversity? (604-606)
10. How does species abundance differ from species diversity? (607)
11. Distinguish between an edge and an ecotone. (608)
12. What is the edge effect? (609)
13. What are some of the characteristics of the edge environment? (609-611)
14. What is the island biogeography theory? (611-613)
15. What is habitat fragmentation? How does it relate to the island biogeography theory? (614)
16. Distinguish between edge and interior conditions. (616-618)
17. What are corridors? (619)
18. What are assembly rules? (619-620)
19. How does competition, predation, parasites, and disease influence community structure? (620-624)
20. What are the major differences between organismic and individualistic concepts of community structure? (625-626)
21. What is the difference between a gradient and a continuum? (627)
22. Distinguish between the continuum index, gradient analysis, community ordination, and principle components analysis methods of community classification? (628-629; Appendix, 723-729)

Key Terms and Phrases

community	species evenness	edge
guild	species diversity	faunal collapse
perennating tissue	species abundance	faunal relaxation
vertical stratification	random niche model	corridor
horizontal structure	niche-preemption hypothesis	assembly rules
dominants	log-normal hypothesis	continuum
keystone species	area-sensitive species	fidelity
relative abundance	inherent edges	constancy
relative dominance	induced edge	ordination
relative frequency	edge effect	continuum index
importance value	turnover rate	gradient analysis

index species   community ordination   rescue effect
species richness   principle components analysis interior species
area-insensitive species gradient   frequency
relative density   ecotone

Key Term Review

1. Groups of species that feed in a similar manner in a given habitat make up a
   _____.

2. _____ _____ are those whose presence is critical to the integrity of
   the community.

3. A naturally occurring assemblage of plants and animals living in the same
   environment and interacting in some way is a _____.

4. The embryonic or meristemic tissues of buds, bulbs, tubers, roots, and seeds are
   _____ _____.

5. The layering of vegetation in a community that affects the physical and biological
   aspects of a community is its _____ _____.

6. Species or groups of species that control the nature of a community are called
   _____.

7. The degree of patchiness of vegetation across a community defines its
   _____ _____.

8. The number of individuals of one species relative to the number of all individuals of
   all species in a community is a measure of _____
   _____.

9. The ratio of total basal area occupied by one species to the total basal area of all
   species is a measure of _____ _____.

10. Number of sample plots in which one species occurs relative to the total number of
    sample plots is a measure of its _____.

11. The frequency value of one species compared to frequency value of all species is a
    measure of _____ _____.

12. The sum of relative density, relative frequency, and relative dominance give the
    _____ _____.

13. Species that have a high importance value in a community often serve as an
    _____ _____ of that community.

14. A number of species in a community is a measure of its _____ _____.

15. The relative abundance of individuals among the species is a measure of
    _____ _____.

16. Species richness and species evenness considered together measure
    _____ _____.

17. _____ _____ measures the manner in which a species divide
    environmental space.

18. The boundary where two or more vegetation types meet is an _____.

19. The rate at which one species on an island is lost and a replacement gained is the
    _____ _____.

20. An edge created and maintained by a disturbance is an _____ _____.

21. When an impending extinction of a dwindling population is slowed or stopped by an
    influx of immigrants it is called a _____ _____.

22. Stable and permanent edges created by abrupt environmental changes are
    _____ _____.

23. The fact that variety and density of life is greatest about edges is called
    _____ _____.

24. The blending of two or more vegetation types along edges produces an
    _____.

25. Species whose habitat begins some distance within a large area of habitat type are
    considered _____ _____.

26. When the number of species in a newly created habitat fragment declines to a new
    lower stable state, the habitat has experienced _____
    _____.

27. If the habitat fragment continues to decrease it will experience _____
    _____.

28. Species that require large territories or foraging areas but are not interior species
    are _____ _____.

29. Species that are at home in both large and small units of habitat are called
    _____ _____.

30. _____ are strips of similar vegetation that connect habitat
    fragments.

31. Mechanisms of how species fit together in a community are called _____
    _____.

236

32. A sequence of communities showing a gradual changes in species is a
    _____.

33. A change of species in response to changing environmental conditions is a
    _____.

34. The persistent appearance of a species in a particular community type is
    _____.

35. The ratio of the species always associated with the community type to the total
    number of species is called _____.

36. An arrangement of communities along a linear axis according to their similarity is
    _____.

37. A synthetic index ordering species relative to changing vegetation composition is
    called a _____ _____.

38. The _____ _____ _____ views species abundance as
    a random partitioning of resources distributed along a continuum.

39. The _____ _____ _____ holds that the
    most dominant species claims the most space and the least successful species the
    smallest amount of space.

40. _____ _____ plots the pattern of plant responses to changes
    along a particular environmental gradient.

41. The _____ _____ presumes that niche space
    occupied by a species is determined by a number of variables, such as food and
    microclimate.

42. In _____ _____ species are arranged to one or
    more ecological gradients; a form of indirect gradient analysis.

43. Complex multivariate methods are involved in the _____
    _____ _____ of communities.

237

Self Test

True and False

1. _____ Herbivory affects the spatial pattern of vegetation.

2. _____ Communities have emergent properties.

3. _____ Keystone species are critical to the integrity of the community.

4. _____ Species diversity considers only the number of species.

5. _____ Tree diversity is greatest in southeastern United States.

6. _____ Vertical stratification of life in aquatic communities is affected by gradients of light, temperature, and oxygen.

7. _____ The random niche model produces the lowest evenness of the three species abundance models.

8. _____ Induced edges are relatively permanent.

9. _____ The variety of species and the density of life are greatest about edges.

10. _____ South-facing edges are drier and warmer than north-facing edges.

11. _____ The transition zone between two vegetation types is an edge.

12. _____ At some point in an area, edge becomes redundant.

13. _____ Interior species decline as their habitat becomes fragmented.

14. _____ Therophytes are commonly found in tropical rain forests.

15. _____ The number of species in a community is species diversity.

16. _____ With more species on an island, extinction rates increase.

17. _____ Keystone species must be predators.

18. _____ The density of edge species varies as a constant proportion of available edge.

19. _____ By controlling competitive interactions among plants, herbivory increases species richness.

20. _____ Equability in island biogeography deals only with the number of species.

21. _____ Disease and parasites have changed the composition of many communities.

22. _____ Ubiquitous species have no strong affinity for any type of habitat.

23. _____ Turnover rates are greater for islands distant from an immigrant pool than for near islands.

24. _____ Edges attract specialist species.

Matching

Which of the folowing characteristics apply to edge vegetation or to edge animals? Indicate edge characteristics by placing an **E** before the statement and nonedge characteristics with an **N**.

1. _____ Shade tolerant plants.

2. _____ Ability to disperse widely.

3. _____ Colonize disturbed sites.

4. _____ Plants have relatively long life cycles.

5. _____ Habitat requirements involve one or more adjacent vegetation patches.

6. _____ Species are opportunistic.

7. _____ Plants are moisture-demanding species.

8. _____ Non-persistent habitats.

Associate each of the Raunkiaer life forms with their appropriate description:

A. Hemicryptophytes   C. Geophytes   E. Epiphytes
B. Therophytes   D. Phanerophytes   F. Chamaephytes

9. _____ Survives unfavorable periods as seeds.

10. _____ Perennating shoots or buds range from the surface of the ground to about 25 cm above the surface.

11. _____ Perennial buds are carried well up in the air.

12. _____ Grows on other plants with roots in the air.

13. _____ Buds are buried in the ground on bulbs or rhizomes.

14. _____ Buds or perennial shoots are close to the surface; often buried in litter.

A number of hypotheses have been proposed to explain latitudinal variations in species diversity. Associate each of the hypotheses with its appropriate descriptive statement.

A. Evolutionary time hypothesis
B. Spatial heterogeneity hypothesis
C. Climate stability hypothesis
D. Stability-time hypothesis
E. Productivity hypothesis

15. _____ A stable climate provides a more favorable environment, encouraging a high species richness.

16. _____ Level of diversity is determined by the amount of energy flowing through a food web.

17. _____ The more complex the environment, the greater the species diversity.

18. _____ Regions with fluctuating physically controlled environments, subjecting species to physiological stress, has lower species diversity than regions with uniform physical conditions that favor biologically controlled environments.

19. _____ Evolutionarily old communities have higher species diversity than younger communities that have not had time for new species to evolve.

Labeling

Draw a simple diagram using two curves to show the relationship between immigration rates, extinction rates, and species equilibrium in an island situation. Label all axes and equilibrium point.

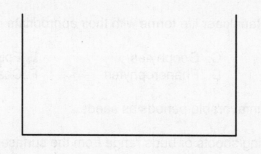

Short Answer Questions

1. How do terrestrial habitat islands differ from oceanic islands?

2. One means of increasing species richness and diversity is to increase the amount of edge in an area. Although edge may increase diversity, what major problems does it create for many species, including those of the edge?

3. How can predation influence community structure?

CHAPTER 29
DISTURBANCE

Chapter Outline

Characteristics of Disturbance
  Intensity
  Frequency
  Scale
Sources of Disturbance
  Fire
  Wind
  Moving Water
  Drought
  Animals

Timber Harvesting
Cultivation
Surface Mining
Effects on Nutrient Cycling
Animal Response to Disturbance
Community Stability
  Equilibrium Communities
  Nonequilibrium Communities
Summary

Learning Objectives

After completing this chapter you should be able to:
- Define disturbance.
- Describe the meaning of intensity, frequency, and scale of disturbances.
- Discuss the role and effects of fire as a disturbance.
- Describe the effects of disturbance by wind, moving water, and drought on ecosystems.
- Discuss the role of herbivorous animals as an influence on the nature of ecosystems.
- Describe the impact of logging on forests and the various methods employed.
- Explain why cultivation and surface mining are such severe disturbances.
- Tell how major disturbances effect nutrient cycling in ecosystems.
- Discuss how animals respond to disturbances, especially fire.
- Distinguish between equilibrium and nonequilibrium communities and between local and global stability.
- Explain the intermediate disturbance hypothesis.

Summary

After reading this chapter and before continuing with the following material, read the Chapter Summary on pages 653-654. You will find related material on the following pages: plant response to drought, 72-73; nutrient cycling, 116-120; r and K selection, 448; grazing and grasslands, 234; herbivory and community structure, 622; predation and community structure, 621-623; parasitism and community structure, 623-624; secondary succession, 659-660; cyclic succession, 666-667.

Study Questions

1. Define disturbance. (634)
2. Distinguish between intensity, frequency, and scale of disturbance. (634-641)
3. What effect can suppression of a natural disturbance have on a disturbance-controlled ecosystem? (638)
4. What is gap formation? How does it relate to the scale of disturbance? (638)
5. What environmental conditions change within a gap and how does vegetation respond? (638)
6. What constitutes large scale disturbances relative to small scale disturbances? (641)
7. How do woodland herbs appear to respond to canopy removal? (641-642, 647)
8. Why is fire such an important ecological disturbance? (641)
9. What are the three types of forest fires? How do they differ relative to fire intensity? (641-642)
10. What are some adaptations of vegetation to fire? What are fire adapted ecosystems? (642-643)
11. What influences the frequency of fires? What is the ecological effect of infrequent fires? Too frequent fires? (635-638)
12. What ecological effects does wind have on a vegetation community, especially forests? (644)
13. How does drought act as a disturbance? (645)
14. How does disturbance by herbivores impact forest and grassland ecosystems? (645-646)
15. Distinguish between clearcutting, strip cutting, shelterwood cutting, and selection cutting? What are the ecological effects of each on a forest? (646)
16. How does forest vegetation respond to the sudden removal of a forest canopy? (647)
17. What is the effect of the combined disturbances of grazing, fire, and drought on community structure? (645-646)
18. What major differences exist between cultivated and natural ecosystems? (647-648)
19. What impacts does surface mining have on the environment? (648)
20. How do fire, timber harvesting, and cultivation affect nutrient cycling in those disturbed ecosystems? (649-651)
21. What effects do various small-scale and large-scale disturbances have on animal life? (657)
22. How does animal life respond to fire? What are fire-dependent species? (651)
23. What role does disturbance play in ecosystem stability? (652-653)
24. Define stability, local stability, and global stability. (652-653)
25. Distinguish between resistance and resilience. (652)
26. Distinguish between equilibrium and nonequilibrium communities. (652-653)
27. What is the intermediate disturbance hypothesis? (653)

Key Terms and Phrases

disturbance	ground fire	resilience
intensity (of a disturbance)	serotiny	resistance
frequency	selection cutting	local stability
scale	even-aged management	global stability

return interval   clearcutting   equilibrium community
gap   strip cutting   nonequilibrium community
crown fire   shelterwood cutting
surface fire   intermediate disturbance hypothesis

Key Term Review

1. Any physical force that damages natural systems and results in the mortality of organisms or loss of biomass is a _____.

2. The mean number of disturbances that occur within a particular time interval is _____.

3. An opening created by a small scale disturbance that becomes a local site for vegetation growth and regeneration is a _____.

4. The proportion of the total biomass or a population of a particular species that is killed or removed is a measure of a disturbance's _____.

5. The mean time between disturbances on the same piece of ground is the _____ _____.

6. The range in the areal size of a disturbance is its _____.

7. A fire that sweeps through the canopy of a forest is a _____ _____.

8. A fire that feeds on the litter layer of grasslands and forests is a _____ _____.

9. A fire that consumes organic matter down to the mineral substrate or bare rock is a _____ _____.

10. Cones of certain coniferous trees that open to release their seeds only when exposed to fire is _____.

11. The ability of an ecosystem to withstand major disturbances with little change is a measure of its _____.

12. The speed with which an ecosystem returns to its same general state after a major disturbance is a measure of its _____.

13. The removal of a single or small groups of mature trees from a forest stand is _____ _____.

14. _____ _____ involves the removal of a forest stand and its reversion to an earlier successional stage .

15. The tendency of a system to return to its original state from a small disturbance is _____ _____.

16. The tendency of an ecosystem to return its original condition from all sorts of disturbances is _____ _____.

17. _____ is the removal of all timber from a stand.

18. The removal of all merchantable timber and remaining trees from wide strips , leaving remaining strips for future logging is _____ _____.

19. _____ _____ leaves 10 to 70 percent of a stand after the initial cutting; the remaining trees are not removed until new regenerating growth is well under way.

20. The idea that highest diversity in an ecosystem is maintained at an intermediate level of disturbance is called the _____ _____ _____.

21. An _____ _____ is one that is highly stable and resists departure from that condition.

22. A _____ _____ is highly influenced by disturbances; it fluctuates from high to low diversity depending upon the nature of the disturbances.

Self Test

True and False

1. _____ Most communities are nonequilibrium communities.

2. _____ The tendency of community to reach a steady state is its stability.

3. _____ Nitrogen released from biomass by fires is in a form available to plants.

4. _____ Fire is a major agent of destruction for wildlife.

5. _____ A rapid return of an ecosystem to its same general status after a disturbance is a measure of its resistance.

6. _____ A response of a forest to gap formation is canopy closure.

7. _____ Diversity in a landscape comes about through some form of disturbance.

8. _____ Intensity of a disturbance has little relationship to the frequency of disturbance.

9. _____ A small scale disturbance in a forest creates different microclimates in a forest stand.

10. _____ The return interval of a disturbance relates to the life span of the plant species involved.

11. _____ If the frequency of a disturbance is high, biomass accumulates.

12. _____ A vigorous growth of ferns can adversely affect the growth of forest seedlings.

13. _____ Interference with the normal recovery of an ecosystem from a disturbance can permanently change the nature of the system.

14. _____ Fire had little influence on the development of the eastern deciduous forest of North America.

15. _____ Selection cutting mimics natural gap formation in a forest.

16. _____ Young trees are more vulnerable to wind damage than mature trees.

17. _____ Some vegetation types possess characteristics that enhance the spread of fire.

18. _____ African elephants are a major influence on the development of savanna vegetation.

19. _____ Overbrowsing of forest vegetation by white-tailed deer can lead to rarity or extinction of certain woodland plants.

20. _____ Frass (droppings) of foliage-eating forest canopy insects returns large amounts of nutrients in the leaves to forest litter.

Short Answer Questions

1. What problems are associated with too frequent and too infrequent disturbances in an ecosystem, especially as they relate to fire?

2. Describe some fire adapted traits exhibited by vegetation?

3. Why do communities kept in a moderate state of disequilibrium through periodic disturbances have the highest diversity?

Answers

Key Term Review

1. disturbance
2. frequency
3. gap
4. intensity
5. return interval
6. scale
7. crown fire
8. surface fire
9. ground fire
10. serotiny
11. resistance
12. resilience
13. selection cutting
14. even-aged management
15. local stability
16. global stability
17. clearcutting
18. strip cutting
19. shelterwood cutting
20. intermediate disturbance hypothesis
21. equilibrium community
22. nonequilibrium community

Self Test

True and False

1. T	6. T	11. F	16. F
2. T	7. T	12. T	17. T
3. F	8. F	13. T	18. T
4. F	9. T	14. F	19. T
5. F	10. T	15. T	20. T

Short Answer Questions

1. Too frequent disturbances result in the degradation of the system. Frequent fires do not allow the original vegetation to recover, grow, and produce seeds. As a result the community may revert to some other type. Too infrequent disturbances, especially in a forest, allow an accumulation of litter and debris that can support an intense fire, senescence and death of trees, and development of an understory that in some forests that can carry a fire to the canopy. These conditions, especially if accompanied by drought, can result in extensive forest fires.

2. Plants of fire-adapted ecosystems have evolved traits that permit protection from or response to fire. Such traits include thick bark that insulates the cambium layer from heat, an accumulation of needles that encourages light periodic surface fires, and rapid growth to get the canopy above the level of surface fires. Some fire-dominated systems use the death of mature plants as a means of regenerating the system by releasing seeds. Other systems rely on fire-stimulated germination of seeds in the soil. Another response is resprouting from buds on roots, rhizomes, root collars, and lignotubers.

3. At a high state of disturbance *r*-selected species that are short-lived and able to colonize disturbed sites dominate the community. At a low state of disturbance, *K*-selected, long-lived species dominate the community. Between these two extremes, however, there is a length of time between disturbances at which species of varying rates of growth, longevity, and competitive ability are able to persist, adding greatly to the diversity of the system.

CHAPTER 30
SUCCESSION

Learning Objectives

After completing this chapter you should be able to:
- Define succession.
- Distinguish between primary and secondary succession and describe an example of each.
- Describe aquatic succession and succession on intertidal zones.
- Discuss the various models developed to explain succession.
- Explain the concept of the climax and its validity.
- Discuss cyclic succession and the shifting mosaic concept as they relate to the climax theory.
- Explain why fluctuations in vegetation are nonsuccessional.
- Describe changes in community attributes through succession.
- Discuss the role of time in succession.
- Explain how succession of vegetation influences succession of animal life.
- Describe degradative succession.
- Discuss how the study of paleosuccession relates to vegetation patterns and changes observed today.

Summary

After reading this chapter and before continuing with the material below, read the Chapter Summary on pages 686-686, You will find related material on the following pages: shade tolerance, 102-104; nutrient cycling, 116-120; nutrient budgets, 118; decomposition, 159-165; rocky intertidal shores, 337-340; competition in plants, 400, 484, 490; *r* and *K* selection, 448; differential resource utilization, 492-493; vertical structure of communities, 599-601; patch dynamics, 638-641; disturbance by animals, 645-646.

Study Questions

1. Define succession. (656)
2. Distinguish between primary and secondary succession. (656-660)
3. Describe the basics of aquatic succession and intertidal succession. (660)
4. What is meant by relay floristic and initial floristic composition as applied to succession? (661)
5. What is the major premise of the ecosystem or holistic concept of succession? (661)
6. What is the major premise of the population or reductionist concept of succession? (661-662)
7. What are the facilitation, tolerance, and inhibition models of succession? (662)
8. What are the major differences between the resource ratio model of succession and the individual-based plant model? (663-664)
9. What is climax vegetation? What is the weakness of the concept? (664-666)
10. What is cyclic succession? (666)
11. What is the shifting mosaic concept and how does it relate to the climax concept? (667-668)
12. What are fluctuations relative to succession? Wave generation? (668)
13. What are some changes in ecosystem attributes during succession? Do observed changes in succession match all of these generalizations? (667-671)
14. How does time enter into succession? (671-672)
15. Is succession always directional? (672)
16. What is the relationship of animals to plant succession? (672)
17. What distinguishes degradative succession from succession? (676)
18. What is paleoecology and how does it relate to present day vegetation patterns and changes? (677)
19. What vegetation changes took place during the Pleistocene? (677)
20. What major vegetation changes took place during the glacial and interglacial periods? (677-683)
21. Of what significance does the study of plant migration in the postglacial period have for the study of current vegetation patterns and potential changes resulting from global climate change? (683)

Key Terms and Phrases

ecological succession	competition	polyclimax theory
sere	reaction	climax pattern theory
seral stage	stabilization	paleoecology
climax	relay floristics	cyclic succession
primary succession	initial floristics	reciprocal replacement
secondary succession	facilitation model	shifting mosaic steady state
pioneer species	inhibition model	fluctuations
autogenic	tolerance model	regeneration wave
allogenic	resource ratio model	reorganization stage
nudation	steady state	aggradation stage

migration   monoclimax theory   transition stage
ecesis   degradative succession   immigration
individual-based plant model   stable stage

Key Term Review

1. A change in species composition and community function over time is
   _____ _____.

2. Organisms that colonize barren areas and highly disturbed sites are _____
   _____.

3. The sequence of communities in succession make up a _____.

4. A distinct successional community is called a _____ _____.

5. Succession that begins on areas unoccupied or unchanged by organisms is
   _____ _____.

6. The end point of succession, considered as a relatively self-sustaining seral stage,
   has been called the _____.

7. Succession that is self-driven, brought about by the organisms themselves, is
   _____.

8. Succession that proceeds on areas where vegetation cover has been disturbed is
   _____ _____.

9. Environmentally-induced succession is _____.

10. According to the Clemensian view of succession, the development of a bare site
    from which succession begins is _____.

11. The arrival of seeds or other propagules to the bare site is _____.

12. The replacement of one plant community by another resulting in the persistence of
    one community, Clements called _____.

13. The establishment and initial growth of vegetation was called _____.

14. The phase of succession when species compete for space, light, and nutrients is
    _____.

15. Clements called the self-induced effects of plants on their habitat _____.

16. The _____ _____ concept considers succession as groups of
    associated species appearing and disappearing together through time.

253

17. The _____ _____ concept of succession proposes that the propagules of most species, both early and late successional, are on the site from the start.

18. The _____ _____ of succession proposed by Clements, holds that only one climax exists for a region, whose characteristics are determined solely by climate.

19. Successional changes in which each seral stage is related to others in an orderly upgrade and downgrade series is _____ _____.

20. The _____ _____ proposes that the climax vegetation of a region consists of a mosaic of climaxes controlled by a variety of environmental conditions.

21. Small scale cyclic succession that most frequently occurs in forest gaps is referred to as _____ _____.

22. The _____ _____ describes succession as wholly autogenic, brought about from within by the organisms themselves.

23. The _____ _____ _____ proposes that the total environment of an ecosystem determines the composition, species structure, and balance of a climax.

24. _____ _____ is wholly heterotrophic with maximum energy at the start, followed by a steady decline as succession proceeds.

25. Forest recovery after clearcutting or some other major disturbance starts with a _____ _____.

26. Following this stage is an _____ _____ in which biomass accumulates.

27. Forest growth begins to slow in the _____ _____.

28. Finally the forest arrives at a mature stage or _____ _____.

29. The study of the relationships of ancient flora and fauna is _____.

30. According to the _____ _____ the site belongs to those species that become established first and hold their position.

31. In a _____ _____ _____ the standing crop of living and dead biomass fluctuates about a mean between high net production and biomass accumulation and senescence where respiration exceeds production.

32. _____ are short term, reversible changes in vegetation involving alternation of species within a stand or replacement of one age class by another.

254

33. When trees die off in stages starting at the edge of a disturbance and are replaced by a vigorous young stand of the same species, the phenomenon is called a _____ _____.

34. The _____ _____ suggests that late successional species are neither aided nor inhibited by species of earlier stages

35. According to the _____ _____ _____, succession comes about as the relative availability of resources, particularly soil nutrients and light , changes through time, changing the competitive relationships among species.

36. The _____ _____ _____ proposes that succession is a population process involving competition for resources among individual plants, independent of species. The ability of a plant to compete is restrained by life history traits.

Self Test

True and False

1. _____ Succession takes place as availability of resources changes with time.

2. _____ Plants in early successional stages are large and slow-growing.

3. _____ Each successional community or seral stage is a one-time event not to be repeated.

4. _____ Even in so-called climax communities, succession never ceases.

5. _____ According to the holistic approach to succession, each successional stage alters the environment allowing later stages to become established.

6. _____ Predation has little influence on succession in the rocky intertidal zone.

7. _____ The ultimate winners in succession are long-lived plants.

8. _____ Plants of later successional stages become dominant because they are efficient at exploiting resources.

9. _____ Secondary succession can start with intolerant woody species.

10. _____ Early colonists lose temporary dominance because of differences in competitive ability.

11. _____ Aquatic succession is secondary.

12. _____ In the reorganization stage of forest succession, the ratio of production to respiration is less than one.

13. _____ Early colonizing species are adapted to a high soil nutrient, high light regime.

14. _____ Grazing and browsing herbivores influence succession.

15. _____ Aquatic succession is largely allogenic rather than autogenic.

16. _____ According to the Clemesian view of succession, the climax reproduces itself.

17. _____ Once a forest is cut, the original composition of species will be duplicated by succession.

18. _____ A knowledge of previous vegetation on a site can predict the composition of a forest that will return to an abandoned field.

19. _____ Various patterns of successional change result from different combinations of physiological traits of plants.

20. _____ Shade intolerant trees always replace shrubs in a succession a sequence.

21. _____ Secondary succession takes place on glacial till and sand dunes.

22. _____ Disturbance is necessary to maintain diversity in a "climax" forest.

23. _____ Shrub stages of succession support a higher species diversity than young forest stands.

24. _____ A landscape with various seral stages holds the greatest diversity of animal life.

Matching

Selecting the term that best answers the questions relating to the narrative below. (You will not use all of the terms.)

Through historical studies, field studies, and observations over time, you have learned much about vegetation development on a large tract of forest land. The area was burned over many years ago. After the fire the site was claimed first by fireweed, horseweed, and panic grass, then by aspen, pin cherry, and some pine. In time, those trees gave way to mixed hardwood: red oak, yellow birch, sugar maple, and beech. This forest still occupies the site. As you study the forest, you observe young beech, sugar maple, or yellow birch growing in openings created by blowdown and deaths of canopy trees. Further study and observations lead you to discover that microsuccession involves colonization of openings, first by yellow birch, then by beech, followed by sugar maple, and then back to yellow birch. Throughout the forest, however, sugar maple and beech seem to be maintaining themselves.

A. Opportunistic   E. Fluctuation   I. Shade intolerant
B. Secondary   F. Equilibrium   J. Shade tolerant
C. Stable   G. Facilitation   K. Reciprocal replacement
D. Primary   H. Inhibition

1. _____ Type of succession that has taken place on the site.

2. _____ Aspen and pin cherry belong to what general class of species?

3. _____ Based on the given information, how would you classify the successional process involving aspen and pin cherry?

4. _____ Based on their response to light, sugar maple and beech are considered what?

5. _____ What would you call the yellow birch-beech-sugar maple-yellow birch successional pattern in the forest?

Associate each of the following physiological and life history traits with early **E** or late **L** successional plants.

6. _____ High photosynthetic efficiency at low light intensity.

7. _____ Respiration rate is high.

8. _____ Seed viabiability is long.

9. _____ Seed size is large.

10. _____ Root-to-shoot ratio is low.

11. _____ Transpiration rate is high.

Associate each of the descriptive statements with the appropriate model of succession.

A. Facilitation model   D. Inhibition model
B. Resource ratio model   E. Individual-based plant model
C. Tolerance model

12. _____ A location is colonized by a species of plants that maintain their position against all invaders.

13. _____ Each successional stage alters the environment allowing later stages to become established.

14. _____ A changing ratio of soil nutrients to light leads to the replacement of one plant species by another.

15. _____ Later successional stages become established because their plants are efficient at exploiting resources.

16. _____ Plants both respond to the prevailing environment and influence the availability of resources in the environment.

Short Answer Questions.

1. Why did the impact of the last glaciation on the vegetation differ between North America and Europe?

2. We face the potential of the warming of the global climate. Just as happened at the end of the glacial period, forest vegetation would advance northward toward and into the tundra, except for major obstacles? What are they?

3. The mechanistic models of plant succession are based on three premises to explain species replacement on a spatial and temporal gradient. What are these premises?

ANSWERS

Key Term Review

1. ecological succession
2. pioneer species
3. sere
4. seral stage
5. primary succession
6. climax
7. autogenic
8. secondary succession
9. allogenic
10. nudation
11. immigration
12. stabilization
13. ecesis
14. competition
15. reaction
16. relay floristics
17. initial floristics
18. monoclimax theory
19. cyclic succession
20. polyclimax theory
21. reciprocal replacement
22. facilitation model
23. climax pattern theory
24. degradative succession
25. reorganization stage
26. aggradation stage
27. transition stage
28. stable stage
29. paleoecology
30. inhibition model
31. shifting mosaic steady state
32. fluctuations
33. regeneration wave
34. tolerance model
35. resource ratio model
36. individual-based plant model

Self Test

True and False

1. T	9. T	17. F
2. F	10. T	18. F
3. T	11. F	19. T
4. T	12. T	20. F
5. T	13. F	21. F
6. F	14. T	22. T
7. T	15. T	23. T
8. T	16. T	24. T

Matching

1. B	9. L
2. A	10. E
3. G	11. E
4. J	12. D
5. K	13. A
6. L	14. B
7. E	15. C
8. E	16. E

Short Answer Questions

1. In Europe the southward retreat of vegetation in front of advancing glaciers was halted by alpine glaciers, mountains, deserts, and seas, resulting in decimation of much of the flora, except for boreal species. In North America no such barrier existed and vegetation, especially deciduous forest trees, was able to retreat far south of the glacier.

2. Global climate change poses the threat that current boreal forests would move northward into the tundra, which would mostly disappear. As in postglacial times, deciduous forests should also advance northward, except for two major obstacles, physical and temporal. The landscape of eastern North America is so fragmented by megalopolises and urban and suburban developments that deciduous tree species would encounter serious obstacles to northward dispersal. Further, the rate of warming is such that tree species would have to advance at the rate of five to ten kilometers a years compared to fractions of a kilometer a year they moved as the glacier receded.

3. The mechanistic models of succession are based on three premises. One premise is that as plants grow they alter the environment. This alteration changes the availability of resources and at the same time the competitive relationships among plants. The second premise is that physiological traits of various plant species are such that their competitive abilities increase or decrease as environmental conditions, especially nutrients and light, change. The third premise is that because of the interaction of premises one and two, species that are good competitors under one set of environmental conditions are poor competitors under another. The changes in competitive abilities influence the direction and nature of succession.